The Aquatic Snails
of the
Family Hydrobiidae
of
Peninsular Florida

The Aquatic Snails of the Family Hydrobiidae of Peninsular Florida

By Fred G. Thompson

University of Florida Press / Gainesville / 1968

A University of Florida Press Book

Library of Congress Catalog Card Number 68-9707

PRINTED FOR THE PUBLISHER BY THE
H. & W. B. DREW COMPANY
JACKSONVILLE, FLORIDA

Introduction

THIS IS a study of the zoogeography, systematics, and ecology of the fresh and brackish water snails of the family Hydrobiidae that inhabit peninsular Florida. The geographic limits of this study include all of the Florida peninsula from the Suwanee River system and the St. Marys River system south, and all of the Florida Keys. These limits were chosen because the molluscan fauna of the peninsula is different from that to the north and west, and the important transition occurs in the region of the neck of the peninsula.

The purpose of this study is to provide a sound basis for classification of the members of the family Hydrobiidae that inhabit the Florida peninsula. Thirty-five species representing ten genera are treated. Twenty-three species and five genera are described as new. A classification based on the morphology of the shell and the soft anatomy is proposed. The geographical and ecological distributions of all species are discussed.

The geographical and ecological distributions of most of the species suggest that the peninsula of Florida was maintained as an emerged peninsula since the Pliocene, and was not reduced to an island or archipelago as some biogeographers and geologists have proposed. This interpretation is necessary because the aquatic environments needed to maintain the characteristic Florida entities through the Pleistocene would have been eliminated had the sea level been raised to the extent necessary to submerge the neck of the peninsula. Ecological and geographical considerations argue against the possibility that these snails invaded the peninsula from the north since the proposed inundation.

This study was initiated in September, 1961, and was supported in part by a National Science Foundation Institutional Grant, and in part by funds provided by the Department of Zoology, University of Miami. Final preparation of the manuscript was completed under the National Institutes of Health Research Grant GM 12300. The study is based on material collected in Florida during 1961 through 1964, and on observations made in the field during this time. It is supplemented by material borrowed from other museums, and material collected in Mexico and Costa Rica during 1963.

I am grateful to the members of the graduate faculty of the Department of Zoology, University of Miami for the many courtesies and the guidance extended to me during this study. Drs. H. Frederick Strohecker, W. Henry Leigh, Earl R. Rich, and Burton P. Hunt served on my dissertation committee. Dr. Strohecker acted as my graduate advisor during the course of my study. Dr. Henry van der Schalie of the Museum of Zoology, University of Michigan served as a consultant. Dr. Joseph P. E. Morrison of the United States National Museum has aided me with his extensive knowledge of this group of snails.

The following people have loaned me specimens for examination and comparisons from the collections in their charges: Dr. R. Tucker Abbott, Academy of Natural Sciences of Philadelphia, Dr. Harald A. Rehder, United States National Museum, Dr. William J. Clench, Museum of Comparative Zoology, Harvard University, Dr. Henry van der Schalie, Museum of Zoology, University of Michigan, and Dr. Juan Parodiz, Carnegie Museum. I am grateful to the following people, who have aided me in field work, and in many other ways that have led to the completion of this study: Dr. F. Wayne King, Mr. Thomas H. Krakauer, and Dr. Dennis R. Paulson. The skillfully executed drawings of the shells were made by Mrs. Barbara Webb.

Fred G. Thompson

Gainesville, Florida
December 2, 1968

Contents

Figures

Illustrations

Tables

Historical Background

MEMBERS OF the family Hydrobiidae are found in fresh and brackish water throughout the world, with the exception of some remote oceanic islands. Several marine genera are also found in various regions. The family is very large and includes about 100 genera and a multitude of species. About 25 genera and 175 species have been recorded from the United States.

Much disagreement exists concerning the proper name for this family of snails. European authors have generally used the name Hydrobiidae, whereas American authors have generally used the name Amnicolidae. Troschel (1857, 106) first indicated the separation of the group when he proposed the name Hydrobiae, but this name was not defined until 1865 (Stimpson, 4). The name Amnicolidae was proposed and defined by Gill in 1863. Arguments about the use of the name revolve about which takes priority, the date of the name or the date of its definition. This subject has been adequately reviewed by F. C. Baker (1928, 80). The name Hydrobiidae has come into dominant usage during recent years.

The historical development of the study of this family in North America was reviewed by Berry in 1943 (10-13). Since that time four new genera (*Texadina* Abbott and Ladd, 1951; *Notoqillia* Pilsbry, 1953; *Antroselatus* Hubricht, 1963; and *Viascalba* Morrison, 1965) and eight new species have been described from eastern North America (Hubricht, 1957, 1963; Pilsbry, 1950, 1951). Taylor (1966) described five genera endemic to Quatro Cienegas, Coahuila, Mexico. These genera are mostly neotropical in their affinities, although one belongs in the *Somatogyrus* tribe (pp. 99-145).

The classification of the family is in a state of confusion. This is due to the large number of genera and species involved and the cosmopolitan distribution of the family. The vast majority of the species have not been anatomically studied, and the shells of most forms are small, drab, characterless structures that defy identification at the generic level. It is not surprising that most nominate species probably are not identified with their proper genera.

Thiele (1931, 136-59) was the most recent author to attempt a comprehensive classification of the Hydrobiidae. Unfortunately, he was shackled by concepts proposed by earlier authors regarding the validity of shell, opercular, and radular characters, and by a general lack of knowledge of the anatomy of the many forms. Most later authors failed to improve the situation. Berry (1943) studied the Hydrobiidae of Michigan and provided much critical information about the anatomy of the forms in that area, but he failed to utilize new information where it contradicted the classifications established by previous authors, particularly with regard to subfamilies and genera. Morrison (1949, 13-15) proposed a classification of the American members of the family based on the presence of various ducts in the male copulatory organ (the verge). His system approaches a natural arrangement more closely than any other yet proposed, and three of his subfamilies probably are natural assemblages (Bythinellinae, Buliminae, and Emmericiinae). His fourth subfamily (Hydrobiinae) is based on the presence of a single simple duct (the vas deferens) in the verge. The concept of this subfamily does not allow for convergences due to the loss of other ducts, or divergence through other fundamental characters. The Hydrobiinae is the largest of the subfamilies that he recognized, and many of its forms must be studied before it can be determined whether it is or is not a natural group. For purposes of convenience I am following Morrison's system, and accept the Hydrobiinae as a natural division of the family. The Bythinellinae is equivalent to the Amnicolinae of most American authors.

Taylor (1966) recently studied the aquatic mollusks of the Quatro Cienegas Basin of Coahuila, Mexico. In this study he divided the Hydrobiidae into eleven subfamilies, three of which were endemic to the basin. My comments on this work are restricted to matters pertinent to classification of the Florida fauna. *Lyogyrus* (type of the Lyogyrinae) is only subgenerically distinct from *Amnicola* (type of the Amnicolinae). The only basis for recognition of the Lyogyrinae is the degree of coiling of the operculum, which is found to be highly variable in three Florida genera (*Amnicola, Cincinnatia,* and *Aphaostracon* new genus). Thus, the Lyogyrinae cannot be considered distinct from the Amnicolinae.

Cochliopa (type of the Cochliopinae), *Mexithauma* (type of the Mexithauminae), and *Aroapyrgus* (placed in the Hydrobiinae) are similar anatomically to the extent that separation depends on shell characters. They are also similar in having a cylindrical verge with a constricted, slender, terminal penis (Baker, 1930; Morrison, 1946; Taylor, 1966). Some genera of the Cochliopinae (*Cochliopa* and *Subcochliopa*) may also have secondary papillae on the verge (Morrison, 1946), a character that would place them in the Hydrobiinae as defined by Taylor. Shell characters used to define the Cochliopinae (trochoid to planispiral, narrowly to broadly umbilicate) also occur in the Hydrobiinae *(Birgella, Gillia, Somatogyrus,* and so forth). Shell characters used to define the Mexithauminae (periostracal bristles) occur in similar form in the hydrobiin genera *Littoridinops* and *Spilochlamys* new genus. Thus, the recognition of the Cochliopinae and the Mexithauminae becomes highly subjective, and at best fulfills the needs only of those who are concerned with identifications as opposed to those who are concerned with phylogeny.

The genus *Nymphophilus* (type of the Nymphophilinae) is closely allied in all of its characters to the genera that I place in the *Somatogyrus* tribe, and which Taylor places in the Hydrobiinae. Taylor defined the Nymphophilinae on the basis of the shape and structure of the verge, the presence of a blood sinus in the verge, and a raised ridge enclosing the vas deferens on the nape of the male. The shape and structure of the verge are basically identical to those described below for *Cincinnatia, Notogillia,* and *Spilochlamys.* The structure that Taylor identifies as a blood sinus is actually a multiganglionated nerve chord as described below for *Littoridinops,* and as occurs in the verge of all hydrobiids that I have studied. The ridge formed over the vas deferens on the nape of males differs only in degree from the condition that exists in *Cincinnatia* and most other genera. Thus, the Nymphophilinae possess no characters that distinguish them from the Hydrobiinae.

The subfamily Littoridininae is a heterogenous assemblage of genera that is divisible into two or more groups. One group (*Littoridinops, Pyrgophorus,* and *Mexipyrgus*) has *nonglandular* papillae along the margins of the verge (as in the *Hydrobia* tribe, p. 35). The second group (*Texadina*, and *Heleobia* and its allies from Lake Titicaca) has lateral glands along the verge (as in the *Heleobia* tribe, p. 20). Neither group shows significant departure from the anatomical diversity that occurs among the genera that Taylor placed in the Hydrobiinae.

The remaining subfamilies that Taylor recognized are characterized to varying minor extents. At the most I feel that they should be retained as tribes within the Hydrobiinae.

On the basis of our current knowledge of the anatomy of the Hydrobiidae the system of classification proposed by Morrison (1949) is the most satisfactory that has been presented. Its categories are more objectively based than equivalent categories proposed by other authors. Further division of his Hydrobiinae into additional subfamilies may be necessary, but it cannot be done with our present knowledge of hydrobiid anatomy.

No previous comprehensive study of the Florida Hydrobiidae has been attempted. What is known about the fauna has come about through small, scattered, incidental reports. Frauenfeld (1863) was the first investigator to record members of this family from Florida. His paper included the description of *Littoridinops monroensis* and *Cincinnatia floridana*. He also recorded *Amnicola porata* Say and *Flumnicola nuttaliana* (Lea) from the state. His record of *A. porata* probably was based on a misidentification of *Amnicola johnsoni* Pilsbry, and *Flumnicola nuttaliana* probably was based on a misidentification of *Notogillia wetherbyi* (Dall). Dall (1885) added to the list with his description of *Notogillia wetherbyi* and *Cincinnatia monroensis*, and gave additional comments on the identity of *Cincinnatia floridana*. In a series of short papers Pilsbry added the following species to the faunal list of Florida: *Hyalopyrgus aequicostatus* (1889), *Hyalopyrgus brevissimus* (1890), *Amnicola dalli* Pilsbry and Beecher (1892), *Amnicola johnsoni* (1899), *Aphaostracon monas* (1899), *Amnicola sanctijohannis* (1899), *Amnicola augustina* (1904), *Amnicola sablensis* (1951), *Fontigens oxybelis* (1950), and *Fontigens weberi* (1950). *Amnicola sanctijohannis* and *Amnicola augustina* are synonyms of *Cincinnatia floridana; Amnicola sablensis* is a synonym of *Littoridinops monroensis; Fontigens weberi* is probably a Pliocene fossil; and *Fontigens oxybelis* is synonymous with *Hyalopyrgus aequicostatus.* Rhoads (1899, 47) first recorded *Pyrgophorus coronatus* (Peiffer) from

Florida; it was a misidentification of *Pyrgophorus platyrachis* (new species). Dall (1913) described *Amnicola harperi* from the Pleistocene marl of Lake Panasoffkee, Florida. This is synonymous with *Hyalopyrgus brevissimus* (Pilsbry). Vanatta (1934) described *Cincinnatia vanhyningi* from a small spring in central Florida.

The only species that has received any study subsequent to its original description is *Notogillia wetherbyi*. Originally described as a species of *Gillia*, it remained in that genus until Pilsbry (1953, 439) established the genus *Notogillia* based on anatomical characteristics. Subsequently, Clench and Turner (1956, 124-25) added much to our knowledge of its distribution. All other species have remained as isolated names in the literature.

Materials and Methods

THIS STUDY was initiated in September, 1961. Field work was conducted from then through April, 1963, and supplementary field trips were made from September, 1963 through April, 1964. Nearly all accessible inland and brackish waterways in the peninsula were visited on one or more occasions. Samples were taken from a total of 481 field collecting stations (Fig. 1). Hydrobiid snails were found in all counties within the peninsula except Baker, Union, and Osceola. Apparently the flatwood swamps in Baker County and Union County are unfavorable for these snails. Osceola County has many large lakes which one would expect to harbor hydrobiids, but no species have been found there.

Snails were collected from nearly all types of aquatic habitats in the state, except flatland marshes and some small landlocked lakes. Specimens were found in a large variety of microhabitats, including aquatic plants, submerged objects, bare sand, and soft muck, and were routinely collected by shaking aquatic plants and other objects in a pail of water, or by sifting sand and silt with a fine meshed screen (about 10 meshes per centimeter).

Most specimens were preserved and dried for their shells. Selected specimens from most field samples were narcotized with menthol crystals for anatomical studies. The relaxed animals were fixed with Bouin's picro-formalin-acetic acid fixative and preserved in 70 per cent ethanol. Verges were removed from males and studied as unstained or as stained whole mounts and in serial sections. Whole mounts and serial sections were stained with Harris' Haematoxylin. Characteristics of the verge were determined from its apparently erect condition when narcotized.

FIGURE 1

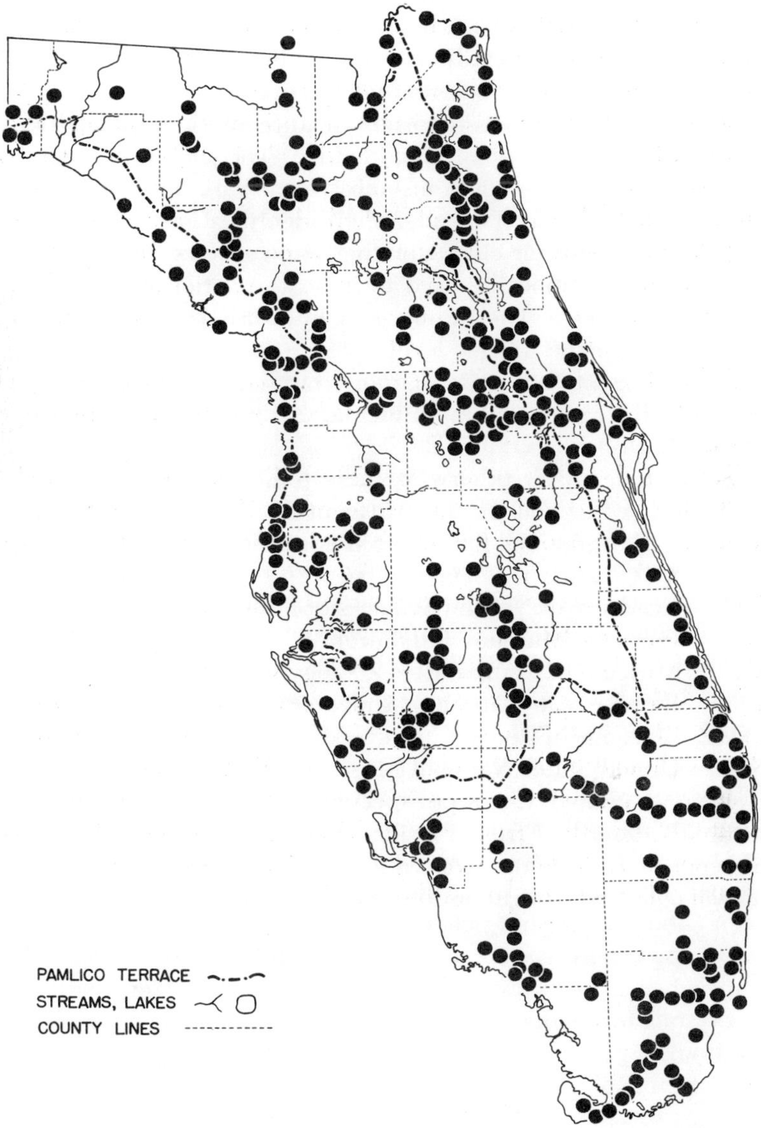

Field stations from which fresh and brackish water mollusks were collected during the course of this study. Many localities are not shown on the map because they lie too close to others to be separable.

The shells were measured with a calibrated Whipple ocular micrometer in a dissecting microscope. The diameter of the initial whorl was measured perpendicular to the initial suture. Measurements of anatomical details and the radula were made with an ocular micrometer in a compound microscope.

The verge is the most important feature of the animal for determining its generic and specific identity. The shell is of secondary importance because in some instances local variations and convergences of shell characters render shell identifications tenuous. The operculum is useful for characterizing some species, but among the Florida forms it cannot be used to indicate relationships at the generic level or above. The radula is of little use as a taxonomic tool at the generic or the specific level.

All drawings of the shell, anatomy, radula, and operculum were made with the aid of a camera lucida. All records are based upon material that I have examined.

All material, unless otherwise indicated, has been deposited in the mollusk collection at the Museum of Zoology, University of Michigan. Designations for museum collections cited in this paper are:

ANSP	Academy of Natural Sciences of Philadelphia
CM	Carnegie Museum, Pittsburgh
MCZ	Museum of Comparative Zoology, Harvard
UF	Florida State Museum, University of Florida, Gainesville
UMMZ	Museum of Zoology, University of Michigan, Ann Arbor
USNM	United States National Museum, Washington, D.C.

I have examined the type specimens of all recent species of hydrobiids described from Florida. I have not examined the original types of *Littoridinops tenuipes,* but I have examined extensive material collected from about the type locality.

General Ecological Considerations

MANY TYPES of aquatic habitats occur throughout the state. The most important of these are sloughs, marshes, rivers, small streams, springs, and lakes. Sloughs are generally devoid of mollusks because of the high concentrations of humic acids. Notable exceptions are sloughs along the lower half of the St. Johns River system. This part of the river system is fed by a multitude of springs, and the surface waters through this region have many of the chemical properties of spring waters.

Of the several types of marshes throughout the peninsula, only salt marshes regularly harbor large hydrobiid populations. Cypress swamps usually have a very sparse molluscan fauna, although occasional abundance may be encountered. The extensive flatwood marshes of Nassau, Duval, Baker, Union, Osceola, Indian River, and Okeechobee Counties are nearly devoid of mollusks.

Of the various lentic situations that occur in Florida, springs and spring runs have the richest and most diversified molluscan fauna. Small streams may also have large molluscan populations due to the great diversity of microhabitats that result from the many pools, small rapids, and alternating marshy and sandy shores that characterize such streams. In larger streams and rivers mollusks tend to be localized and colonial in distribution.

The many lakes throughout the peninsula are highly variable in biological fitness. Some are deep, modified sinkholes in limestone bedrock. Many are shallow depressions with a very gentle shoreline and bottom contour. These have substrates that vary from hard, clean sand to soft ooze or muck. The waters are highly variable in chemical composition, ranging from 3 to 150 ppm in total hardness, and may be clear or very turbid. Most of the lakes have a rich molluscan fauna, but some landlocked lakes in Polk, Orange, and Osceola Counties are devoid of mollusks, except for a few physids and planorbids, due to the absence of an avenue of entry.

Many lakes that have an abundant molluscan fauna originated from sinkholes and presently are landlocked. Some have active springs on the bottom, and others probably have had springs or underground stream connections in the past. The presence of mollusks in these lakes does not necessarily reflect fortuitous colonization, for spring or underground streams could serve as an adequate avenue of entry.

Aquatic operculate snails, particularly the hydrobiids, are bound to their environment and are not easily dispersed over land. It is very unlikely that a hydrobiid could move over land to a new watershed, except by stream capture or by some fortuitous event. Only a few extralimital species leave the water and move about on damp ground, and these have done so through a high degree of specialization in their adaptation to microhabitats. Live snails or their eggs could be carried to new areas by cyclonic winds or in mud on the feet of wading birds, but judging from the distributions of most forms, such introductions are rare. Thus, the present distribution of most species of the family Hydrobiidae probably reflects the geological history of the drainage systems occupied by these snails.

Zoogeographic Relationships

ANY BIOGEOGRAPHIC STUDY is contingent upon a thorough knowledge of the phylogenetic relationships of the forms involved, and is greatly aided by paleontological data. This study is hindered in both of these respects.

Anatomical characters are the only reliable bases for establishing phylogenetic relationships in the family Hydrobiidae. Unfortunately, too little is known about the anatomy of many nearctic and neotropical species to state their proper relationships. Large gaps remain in our knowledge of the hydrobiid fauna of many of the southern states, the gulf coastal area, Mexico, Central America, and the West Indies. Thus, the ranges and diversity of many genera may be much greater than is presently supposed, and apparent faunal similarities or differences between adjacent regions are subject to change with the acquisition of new material.

Although Florida Pliocene and Pleistocene deposits contain large numbers of highly diverse hydrobioid shells, these fossils are useless, for they cannot be identified with any degree of certainty to the family Hydrobiidae, much less to any of the recent genera.

Taylor (1960) used Pleistocene hydrobiids, as well as other mollusks, to demonstrate climatic conditions and faunal shifts of the Pleistocene. The identification of fossil species and corresponding interpretations may be quite safe when one is working with deposits that are relatively young and have a general widespread biotic composition, as with the High Plains region of the United States. When studying fossil faunae that are older than the Pleistocene and have diverse and different compositions from the faunae of present and adjacent regions, the identification of hydrobioid forms becomes highly dubious, and ecological and zoogeographical interpretations based on these identifications are ill-founded.

Eight fossil hydrobioid snails have been described from the Pliocene deposits of Florida. These include *Fontigens plana* (Aldrich), 1911; *Fontigens weberi* Pilsbry, 1950; *Fontigens symmetrica* Pilsbry, 1953; *Fontigens palaea* Pilsbry, 1953; *Fontigens ischna* Pilsbry, 1953; *Amnicola lerodes* Pilsbry, 1953; *Probythinella protera* Pilsbry, 1953; *Notogillia fortaxis* Pilsbry, 1953. All of these, except *Fontigens weberi,* were described from Pliocene deposits that have a predominantly marine fauna. *Fontigens weberi* was described as a recent species from a "bone" specimen from West Lake, Everglades National Park. This species does not occur in the region at present, although shells of this species are common in Pliocene road fill near the lake.

There are no features in the preserved shells that exclude these fossil species from the family Rissoidae, nor are there any features that prove their identity with modern hydrobiid genera. In light of the ecological polymorphism exhibited by some recent species (*Littoridinops monroensis, Cincinnatia floridana,* and others to a lesser extent), it is futile to try to identify these fossil forms with living taxa. There is no way of knowing if similarities are due to phyletic relationships or to convergences from different stocks. Ecological and zoogeographical interpretations based on such identifications would be very tenuous.

The hydrobiid fauna of peninsular Florida appear to have been derived from three sources: (1) the West Indies, (2) Mexico and Central America via the Gulf Coast, and (3) the southeastern states, notably Alabama.

Only a single Florida species, *Heleobops docima,* appears to have been derived from the West Indies. This species is closely related to forms occurring in the Bahama Islands and Haiti.

The genus *Pyrgophorus* entered Florida from Mexico along the Gulf Coast. *Pyrgophorus platyrachis* is more like the *Pyrgophorus spinosus* complex of Texas and Mexico than like the *Pyrgophorus coronatus* complex of the Greater Antilles.

All other Florida hydrobiids have been derived from the southeastern states by two or more routes. The Atlantic and/or Gulf Coastal Plains served as an avenue of entry for species of the subfamily Amnicolinae and for the genera *Littoridinops* and *Onobops.* The species of the subfamily Amnicolinae belong to complexes that lie farther north along the east coast. The genera *Onobops* and *Littoridinops* are known from coastal areas from Maryland and Georgia, respectively, to at least the panhandle of Florida, and they probably occur from New England to Mexico.

The remaining Florida hydrobiids have been derived from the Alabama and Mississippi drainage systems or are old relicts. *Cincinnatia* is widespread throughout North America east of the Rocky Mountains, and is a characteristic component of the Mississippi drainage system and streams entering the Gulf of Mexico from Alabama. Ten species are recognized in Florida, all of which are confined to the peninsula. *Cincinnatia floridana* is generally distributed through the northern half of the peninsula as far north as the Santa Fe River, and as far west as the Suwannee River, and is probably the most generalized species of the genus in the state. It has a wide ecological tolerance and inhabits all types of freshwater habitats; it is even slightly tolerant to brackish water. Three other species are recent derivatives of the *C. floridana* stock, and each is confined to a single spring. Six other species form a second group that is confined to the lower half of the St. Johns drainage system. Five of these species are confined to single springs, and the sixth is very restricted in its distribution along the St. Johns River.

The genus *Spilochlamys* contains two species. One is confined to the middle St. Johns drainage system, and the other has a disjunct distribution along the Gulf Coastal region from Levy County north and west to Jackson County. *Spilochlamys* approaches *Cincinnatia* in shell characters but is anatomically more primitive.

Notogillia is a monotypic genus that occurs in the Florida peninsula and has disjunct populations in the panhandle of the state and in adjacent regions of Alabama and Georgia. The species has existed in the peninsula since the Pliocene (Pilsbry, 1953: 441). *Notogillia* is a relict that is not closely related to other modern genera.

Hyalopyrgus contains two remotely related species, both of which are confined to the north central part of the peninsula. This is another genus that has no close affinities with other genera.

Aphaostracon is a large genus containing nine species, all of which are known only from the Florida peninsula. Most of the species are confined to single springs, and they represent considerable evolutionary diversity. The genus has no relationships with other known American genera.

The number of species that are endemic to the Florida peninsula and their degree of differentiation are very impressive. The most noteworthy are *Cincinnatia* (ten species), *Spilochlamys* (two species), *Hyalopyrgus* (two species), *Notogillia wetherbyi, Aphaostracon* (nine species), and *Amnicola* (three species). In order for most of these species to have differentiated to the degree that has occurred, these forms must have existed in the Ocala Island area for a long time.

Cooke (1939, 1945: 248) has given evidence, based on the interpretation of shorelines, that all but a very small portion of Polk County, Florida, was below sea level during the Aftonian interglacial period of the Pleistocene. This view was generally accepted by Neill (1957, 187) in his review of the biota of Florida. During the period of the Kansas glaciation much of the central part of the Florida peninsula emerged and was connected to the mainland to the north; it has remained that way to varying degrees ever since. During the Aftonian interglacial period the characteristic Florida peninsula biotic elements were supposed to have differentiated on Polk Island; during subsequent emergence of the surrounding land they radiated to occupy the area and differentiated further through local isolations brought about by slight fluctuations of the sea level (Neill, 1957: 187-97).

If we accept the postulate that the major portion of the peninsula was submerged to the extent that was claimed (270 feet), then all of the important freshwater habitats would have been eliminated from Ocala Island. A rise in sea level of 100 feet would be sufficient to separate central Florida from the mainland and to create a waterway through the Suwannee Straits and would have been sufficient to inundate all of the (present) important springs, streams, and lakes of the Florida peninsula. The only water system that could have remained would have been small, isolated, landlocked lakes in sandy soil. Movement into and out of these lakes would have been nearly impossible for organisms that are as bound to an aquatic environment as are hydrobiids. Thus, species of *Cincinnatia, Notogillia, Spilochlamys, Aphaostracon,* and *Hyalopyrgus* would probably have been exterminated if they occupied the peninsula prior to the Aftonian Sea, and would have had to remigrate into the peninsula during or after the Kansas glaciation.

There are five arguments against this assumption. (1) The only species of hydrobiids found in Polk County at the present time is

Amnicola johnsoni. If other species had resided here during the Aftonian interglacial there should be relict populations. (2) If the vagility of the snails was such that they could have immigrated into the peninsula since the Kansas glaciation, they should have been able to invade more southern areas of the peninsula as it became exposed. They should also have been able to move into other areas of the southeastern states. Such a high degree of vagility would also preclude the genetic isolation and speciation that has occurred in *Cincinnatia* and *Aphaostracon.* (3) If the recent hydrobiids had immigrated from the north, they would have had to do so across a strait that extended through the present headwaters of the Suwannee River system and the St. Marys River system and to the highlands of Georgia to the north of the headwaters. This area of Georgia is nearly devoid of mollusks, especially hydrobiids, due to the acidic soils which result from the laterites and igneous bedrock that underlie this region. There is no reason to believe that the area would have been more hospitable to snails in the past. (4) If *Cincinnatia, Hyalopyrgus, Aphaostracon, Spilochlamys,* and *Notogillia,* or ancestral forms, had migrated into the peninsula after the Aftonian Sea, it is not likely that they could have differentiated to the degree that they have in so short a time. (5) If these forms evolved on the mainland and migrated onto the peninsula, why should they not occur in their former ranges? It may be argued that *Notogillia wetherbyi* and *Spilochlamys conica* still do occur in those ranges, but populations of these two species are disjunct and confined to small areas of the panhandle region. They are not generally distributed as one would expect in the case of an animal showing a high degree of vagility and ecological tolerance.

The only way that I can account for the presence of the hydrobiid fauna in the peninsula of Florida is to assume that the central part of the present peninsula consisted of a very large island involving most of the land mass from Polk County north to Marion County, or else to assume that this area was a peninsula. This large insular or peninsular land mass must have persisted since the end of the Miocene or early Pliocene, and throughout this time maintained a large watershed, including lentic situations. These views were also expressed by Dall (1903) and Clench and Turner (1956, 104-5) in their work on Florida mollusks. Data from other fields of study support the idea that a peninsula persisted through the Pleistocene.

The highest marine deposits in Florida that contain Pleistocene fossils occur at an elevation no greater than 25 feet above the present sea level (Cooke, 1945: 248-312). All other deposits that are attributed to Pleistocene times are based on the interpretation of ancient shorelines (Cooke, 1945: 11, 12, 248). Such shorelines could have resulted from differential uplifting, or they may be the remains of lacustrine or loess deposits (Russell, 1957: 427-28).

If Pleistocene seas were high enough to have submerged the neck of the Florida peninsula, then they would have been high enough to have submerged the Colombian Isthmus, the Panamanian Isthmus, and the trans-Nicaraguan Isthmus. There is no geological evidence indicating that this occurred (Schuchert, 1935). The present distribution of the flora and fauna of Central America certainly argues against a Pleistocene submergence of these Central American regions.

It is not a necessary corollary that if continental ice masses were to melt there would be a major rise in sea level (Russell, 1957: 427). The pressure exerted on the earth's mantle by the ice masses is held at an equilibrium by the pressures exerted on the mantle by the magma of the earth. If the pack ice was removed by melting, and resulting water was poured into the oceans, there would be a release of pressure on the mantle from above, and elevation of the continental area would tend to come about due to pressure from below (as is happening presently in the Great Lakes region). The emergence of a large continental land mass could be accommodated by a depression of ocean basins, which might result from an increased water volume.

Systematic Discussion of the Family Hydrobiidae Troschel, 1857

THE FAMILY HYDROBIIDAE includes those taenioglossate prosobranch gastropods that have basocones on the face of the central tooth of the radula. The family primarily inhabits brackish and fresh water, but several marine genera are included. The family is generally characterized as follows (Pilsbry and Bequaert, 1927: 212; F. C. Baker, 1928: 77; Thiele, 1931: 136-37; Connolly, 1939: 569; Berry, 1943: 22). The shell may be any of a variety of shapes, but it is usually small and conical. It is dextral, and the aperture may be complete or incomplete across the parietal wall. The shell may be transparent or opaque, and its surface may be simple or highly ornamented with complicated sculpture. The periostracum usually lacks a color pattern. The operculum is simple, paucispiral, multispiral, or concentric, and may be membranous, chitinous, or calcareous. Its inner surface is usually smooth, and lacks a peg- or wing-like projection. The animal has two slender or stocky tentacles in which the eyes are sessile at their external bases. The otocysts contain a single otolith. The foot is longer than wide, is truncate anteriorly with projecting corners, and is rounded posteriorly. The foot does not extend beyond the snout while in locomotion. The operculigerous lobes are well developed and usually extend over the lateral margins of the foot. The gills are located on the inner surface of the mantle. The left gill is reduced to a small vestige or is absent. The right gill consists of a series of 20-50 laminae which are usually wider than high. This series begins at or near the middle of the mantle collar and extends posteriorly and to the left of the pericardium. The animals are dioecious.

The male copulatory organ (the verge) originates on the middle or the right side of the nape beneath the mantle, and may be simple and tubular, or complicated with accessory ducts, dermal glands, and appendages. In this paper, the functional end of the verge through which the vas deferens terminates is referred to as the penis. This may or may not be distinguishable from the rest of the verge as a small appendage. The female is oviparous or ovoviviparous. The eggs are laid singly or in clusters. Ovoviviparous genera in Florida have embryos in the uterus throughout the year. A free-living veliger stage is lacking since embryonic development is completed within the egg capsule. The denticle formula of the radula is 2-1-1-1-2. The central tooth has one or more basocones on each side of the face, a large mesocone, and three or more ectocones on each side of the reflection. The lateral tooth has a large trapezoidal body with a long shaft that extends laterally. The cusps are usually differentiated into entocones, a single mesocone, and ectocones. The two marginal teeth have simple, undifferentiated cusps on the reflection. The bases of the shafts of the lateral teeth and the marginal teeth insert close together and are remote from the central tooth.

The family Hydrobiidae is closely related to the marine family Rissoidae, but the rissoids differ from the former in several important aspects. The operculum usually has a peg- or wing-like process on the inner surface, the eyes are confined within the base of the tentacles, and accessory tentacles are usually present on the mantle and the foot. The central tooth of the radula has basocones that are confined to the inferior margin. The life cycle includes a free-living veliger stage.

Two subfamilies of the Hydrobiidae occur in Florida, the Amnicolinae and the Hydrobiinae. The subfamily Amnicolinae is characterized by having two ducts within the verge. One is a simple vas deferens that discharges through the tip of the penis, the terminal segment of the verge. The second duct originates from a blind pouch in the haemocoel, and discharges through a large, blunt papilla on the left margin of the verge (Fig. 37, D). The Amnicolinae is represented in Florida only by a single genus and three species.

The subfamily Hydrobiinae is characterized by having a single, simple duct, the vas deferens, within the verge. It discharges through the tip of the penis and has no lateral branches or diverticula. This subfamily is represented in Florida by four tribes.

SUBFAMILY HYDROBIINAE TROSCHEL

This subfamily contains a diverse assemblage of genera. Members of the Hydrobiinae are characterized by having only a single simple duct, the vas deferens, within the verge, but the verge may be highly modified in other respects, including the presence of lobes, appendages, papillae, and dermal glands. When they occur these accessory structures form basic patterns and modifications that characterize large groups of species and genera, and may be used to show phyletic relationships. Four phyletic groups occur in Florida. For purposes of convenience the groups are referred to as tribes, but it is not intended that the names of these tribes shall have any nomenclatorial significance.

The *Heleobia* tribe is represented in Florida by *Heleobops,* and includes the South American genera *Heleobia, Strombopoma, Rhamphopoma, Brachypyrgula, Heligmopoma, Lyrodes,* and the North American genus *Texadina.* The tribe is characterized by possessing large circular apocrine glands along the margins of the verge, and a small, simple terminal papilla at the tip of the verge. When relaxed the verge is recurved to the left within the mantle cavity with its distal end pointing posteriorly.

The *Onobops* tribe is represented in Florida by a single genus, *Onobops,* which is the only North American genus known to belong to this tribe. Several American hydrobiids have been described as species of the genus *Hydrobia,* but their true generic identity still remains to be demonstrated. The tribe is characterized by a simple, cylindrical verge without any accessory appendages, papillae, or glands. When relaxed the verge is coiled counterclockwise beneath the mantle, with its distal end pointing posteriorly.

The *Hydrobia* tribe contains the genera *Pyrgophorus, Littoridinops, Hyalopyrgus,* and *Aphaostracon.* The tribe is confined to the Caribbean region and the Atlantic coastal plains in the United States. The tribe may also include the genus *Littoridina* which occurs along the Pacific coast of South and Central America. The group is characterized by papillae along the margins of the verge. The papillae are sensory structures and contain several radially arranged sinuses that are continuous with the blood passages of the verge. The tip of the verge bears a simple conical papilla. When relaxed the verge is recurved to the left beneath the mantle cavity with its distal end pointing posteriorly.

The *Somatogyrus* tribe includes the genera *Somatogyrus, Birgella, Notogillia, Spilochlamys, Cincinnatia,* and *Nymphophilus.* This tribe is confined to North America east of the Rocky Mountains, and is characterized by a ponderous, dorsoventrally compressed verge that bears a flattened, blade-like penis along the right margin, and a large accessory terminal lobe. The outer and the inner surfaces of the verge and the penis bear several or many glandular tubercles and crests which form a characteristic pattern for each genus. When relaxed the verge is folded upon itself, with its distal end pointing forward and lying over the base.

THE *HELEOBIA* TRIBE

A single genus, *Heleobops,* occurs within the geographical limits of this study. It is most similar to *Heleobia* because of the relatively simple operculum and simple, conical shell. *Heleobops* differs from *Heleobia,* as well as from all other allied genera, by having a bifurcate verge, and by having the glands on the verge depressed flush with the surface and facing the snail's body. The verges of *Heleobia* and allied genera have a simple termination, and the glands along the margins are raised on fleshy pedicels (Hubendick, 1955: 321-22; Morrison, 1963: personal communication). The shell of *Heleobops* is sexually dimorphic in size. This has not been observed in the other genera of the tribe.

Heleobops new genus

Shell sexually dimorphic, female about 1.3-2.0 times the size of the male. Shell medium-sized, elongated, conical, 1.7-2.2 times as long as wide; thin, translucent or opaque, light brown to white in color. Umbilicate or imperforate; columella decidedly thicker than lip, but not callused. Spire conical, straight-sided, 1.1-1.7 times the length of the aperture. Sculpture consisting of distinct incremental striations, which are moderately recurved, and irregular, finer but usually distinct spiral striations. 4.0-5.5 whorls in males; 5.0-7.1 whorls in females; nuclear whorl protruding, 0.25-0.30 mm in diameter; remaining whorls regularly increasing in size, flat-sided to strongly arched. Suture very shallow to deeply impressed. Aperture ovate; upper corner acutely angulate to rounded; inside hyaline; axis of aperture lying at about 20-30° to axis of shell; plane of aperture nearly parallel to shell axis. Peristome complete or incomplete across parietal margin; outer lip reversed sigmoid in lateral profile; thin, fragile, sharp; columellar margin weakly reflected, sharp-edged.

Operculum ovate, thin, chitinous, transparent, amber yellow; columellar and parietal margins nearly straight. Paucispiral, consisting of about two and one-quarter rapidly expanding whorls; nucleus lying at about one-fourth of the distance from the columellar margin to the outer margin, and at about one-third of the distance from the base to the apex. Outer surface with numerous fine incremental striations. Inner surface with a large, spiral, hyaline deposit which forms a small knob on the nucleus.

Radula typically taenioglossate; teeth occurring in about 40-51 transverse rows. Central tooth with 1 long mesocone, and 4-6 ectocones on each side; 1-2 basocones are located on each side on a reflection of the lateral margins. Lateral tooth clavate, with a long lateral shaft; reflection bearing 2-1-4 to 3-1-6 cusps. Inner marginal tooth with 14-20 large cusps; base of shaft slightly dilated below middle. Outer marginal tooth with 40-50 very fine, slender cusps along reflection; base of shaft conspicuously expanded to form a short, flange-like, lateral projection.

Foot relatively small, spatulate; truncate anteriorly, bluntly rounded posteriorly; indented on each side at base of pedicel. Snout and pedicel weakly pigmented with melanophores. Tentacles slightly longer than snout when extended; without dark pigmented bands near their tips. Mantle typically hydrobiid, weakly pigmented with melanophores near collar, or not at all pigmented. Gill composed of a long narrow series of 20-30 lamellae, which are largest near the middle of the series and smallest near the collar. Female oviparous.

The verge is a flattened, slender, tapering structure that originates within the mantle cavity slightly to the right of the median line. When erect it extends along the right side of the animal; when relaxed it is recurved to the left side within the mantle cavity with its distal end pointing posteriorly. Its left margin is broadly expanded at the base. Its right margin bears 2-5 flat apocrine glands along the inner surface facing the body (Fig. 38, A). The distal end of the verge is bifurcate. The right fork, which is the penis, bears a small, slender, terminal papilla. The left fork is subequal in size to the penis, and is histologically similar to the parenchymous tissue of the verge, not vacuolated as in *Littoridinops.* The vas deferens is simple and slender. It enters the base of the verge near the right margin, and terminates in the penis at the base of the papilla.

The glands of the verge are 35-80μ in diameter and are delimited by thickenings of epithelium. The verge is generally covered with a single layer of cuboidal epithelium, but at the periphery of the glands, the epithelium increases in thickness to three cell layers. The bases of the glands are encapsulated with a thin layer of compact, fibrous connective tissue. The glands have a large open crypt on the outside of the verge and consist of a single layer of about 75-100 long, columnar glandular cells, which are about 3-8μ wide by about 10-20μ long (Fig. 38, E). A large, oval nucleus is located near the center or the base of each cell. The cytoplasm contains irregular granular bodies that are tentatively identified as mucopolysaccharides.

Type species: *Heleobops docima* new species.

The name *Heleobops* is derived from the Greek and refers to its likeness to *Heleobia.*

Heleobops, which includes three species, is currently known from the Florida peninsula, Andros Island, and Haiti. Two of these species will be described elsewhere as they are extralimited. *Heleobops docima* differs from the Andros species by its opaque imperforate shell, its flat-sided whorls with a weakly impressed suture, its incomplete peristome with an acutely angulate upper corner, and its thickened columella. The Andros species is like *H. docima* in size and proportions, but has a translucent, openly umbilicate shell, moderately arched whorls with a deeply impressed suture, a complete peristome with a rounded or obtusely angulate upper corner, and a thin columella. The Haitian species is very much like *H. docima,* except that is has 6.0-7.1 whorls in mature female shells.

In Florida, *H. docima* may be confused with *Littoridinops monroensis* with which it is commonly found, and *Pyrgophorus platyrachis. Heleobops docima* is distinguished from *L. monroensis* by its sigmoid outer lip, its narrower, more nearly straight-sided spire, its stronger incremental striations and usually more distinct and more numerous spiral striations, and its narrowly rimate or imperforate umbilicus. *Littoridinops monroensis* has a straight-edged outer lip in lateral profile, the spire is stockier and tends to be convex, the sculpture is very weak, and the umbilicus is usually more openly rimate or perforate. The outer lip in *H. docima* is never thickened, whereas it almost always is in *L. monroensis*. The shell of the smooth form of *P. platyrachis* has a thin columella and a much smaller nuclear whorl than that of *H. docima*.

Heleobops docima **new species**

SHELL (Fig. 48, A-F; Fig 58, A, B). FEMALE: elongate conical; opaque; light grayish white or light brown; shiny, with an oily luster. Imperforate, occasionally very narrowly rimate. Spire 1.2-1.7 times the length of the aperture, conical, straight-sided. Sculpture consisting of numerous close, relatively coarse incremental striations, and much finer and more sparse, irregular spiral striations; sculpture equally developed and continuous from embryonic whorls to aperture. 5.0-6.2 whorls; nuclear whorl protruding, 0.25-0.30 mm in diameter, remaining whorls regularly increasing in size, and regularly descending to the aperture, nearly flat-sided, only slightly arched. Suture usually weakly impressed. Aperture ovate-elliptical; interior hyaline; upper corner acutely angulate; axis of aperture lying at about 20° to axis of shell; plane of aperture nearly parallel to shell axis. Peristome continuous or discontinuous across parietal callus, usually the latter, except in very large specimens; columella conspicuously thicker than outer lip and basal lip, but not callused; outer lip thin, sharp, never thickened by an internal callus; outer lip reversed sigmoid in lateral profile, extending forward at the insertion and at the periphery, slightly but broadly indented between the insertion and the periphery, and slightly recurved below the periphery to the base; basocolumellar margin sharp, slightly reflected; parietal callus usually thin and following curvature of preceding whorl.

Measurements of female shells with five or more whorls: length of shell, 3.1-5.1 mm; diameter of shell, 1.6-2.5 mm; length of aperture, 1.2-2.1 mm; width of aperture, 0.9-1.4 mm; length/diameter, 1.67-2.08; length/aperture, 2.12-2.67; whorls/length, 1.15-1.80.

Measurements of type: (female) length of shell, 4.9 mm; diameter of shell, 2.40 mm; length of aperture, 1.90 mm; width of aperture, 1.40 mm; 6.2 whorls; diameter of nuclear whorl, 0.25 mm.

MALE: The male is conspicuously smaller than the female in all populations, although males from one population may be as large as females with the same number of whorls from another population. Males also tend to have a relatively larger and more ample aperture, and in lateral profile the forward curvature of the lip is more pronounced.

Measurements of males: length of shell, 2.1-3.2 mm; diameter of shell, 1.3-1.7 mm; length of aperture, 1.0-1.6 mm; width of aperture, 0.75-1.00 mm; length/diameter, 1.47-2.07; length/aperture, 1.88-2.54; whorls/length, 1.65-2.00; 4.2-5.5 whorls.

The size of the shell is highly variable from one population to the next, although the general appearance of the shell is rather constant. The population from the quarry on Big Pine Key has a higher than normal frequency of complete peristomes. Some specimens from Collier County and Brevard County have a more deeply impressed suture than is usual, but they occur with specimens that are normal in this respect.

Size variations for six populations are represented in Table 1. These samples were selected to show the extremes of variations between different populations, and have no geographic significance.

Operculum (Fig. 29, A). Sub-obovate, with the right and left margins nearly parallel; outer lip and basal margins rounded; columellar margin and parietal margin nearly straight, forming a blunt angle at their intersection; apex pointed. Partially transparent; translucent only over nuclear region. Inner surface with a hyaline, rugose, subspiral deposit, which is thickest in the nuclear region where it forms a small rounded knob over the center of the nucleus.

Animal (Fig. 26, E, F). The gill contains 22-30 lamellae in a long series that begins near the mantle collar. The filaments are largest near the middle of the series. They become slightly reduced in size posteriorly, and very much reduced anteriorly, especially in the male. The mantle is usually unpigmented, except over each gill filament. Occasionally a small blotch may occur near the collar beside or over the gills. The tip of the snout is white. The sides of the snout are lightly pigmented with melanophores, and there may be a series of about 6 vertical dark bars below the periphery, which tend to coalesce ventrally.

Verge (Fig. 38, A-C). The verge is slender, tapering, and about 0.7-1.0 mm long. The right margin bears 2-5 flat apocrine glands, which lie on the inner surface. Occasionally two or more of the glands may be fused to form one long slender gland. The left side of the verge has a large lateral bulge near the base.

Radula (Fig. 33, A). Twenty specimens examined. Teeth arranged in 42-51 transverse rows. Central tooth about 30μ wide;

TABLE 1.—Measurements and Ratios of Female Shells of *Heleobops docima* from Six Populations. The Measurements are Based Upon Fifteen Specimens Selected from Each Population Sample, and Represent the Largest Specimens in the Samples. Measurements are Expressed in Millimeters

Locality	Length	Diameter	Aperture length	Whorls	Length/Diameter
West Lake, Dade Co.	4.2 - 4.8	2.1 - 2.4	1.7 - 2.0	5.9 - 6.1	1.96 - 2.08
Nine-Mile Pond, Dade Co.	4.3 - 5.1	2.2 - 2.5	1.7 - 2.1	5.5 - 6.2	1.91 - 2.08
Marsh 0.2 mi. w.n.w. US 41 & Fla. 29, Collier Co.	3.1 - 4.2	1.7 - 2.2	1.3 - 1.6	5.2 - 6.0	1.78 - 2.02
Merritt Island, Brevard Co.	2.9 - 3.6	1.7 - 2.0	1.3 - 1.6	5.0 - 5.6	1.61 - 1.95
Cudjoe Key, Monroe Co.	3.1 - 3.7	1.6 - 1.9	1.2 - 1.5	5.4 - 5.7	1.82 - 2.00
Quarry, Big Pine Key, Monroe Co.	3.1 - 3.5	1.6 - 1.9	1.2 - 1.5	5.3 - 5.7	1.67 - 2.06

cusp formula varies from $\frac{4\text{-}1\text{-}4}{1\text{-}1}$ to $\frac{6\text{-}1\text{-}6}{2\text{-}2}$. Lateral tooth with 2-1-4 to 3-1-6 cusps. Inner marginal tooth with 14-20 cusps. Outer marginal tooth with about 45 very slender cusps, all of which are too small to be clearly distinguished under oil immersion at a magnification of 1000 X.

TYPE. *USNM* 636131: Nine-Mile Pond, Everglades National Park, Dade County, Florida; collected November 16, 1962 by Fred G. Thompson. Paratypes: *UMMZ* 216491 (200), *MCZ* 258004 (35), *ANSP* 301543 (30), *CM* 65-11 (35).

DISTRIBUTION (Fig. 2). This species is known only from coastal regions of the Florida peninsula and the Florida Keys, and is confined to brackish water. It is common throughout the southern tip of the peninsula and the lower keys, and isolated colonies have been found in Brevard County and Citrus County. It may occur in intervening brackish areas, but standard collecting procedures have failed to indicate its presence in such places.

LOCALITIES. Florida: *Brevard Co.*—Titusville (*MCZ*); Merritt Island, pond 2 mi. e. Titusville; pond 0.4 mi. n. x-section Fla. 406 and U.S. A1A; pond nr. x-section Fla. 402 and Fla. 406. *Citrus Co.*—Homosassa Springs. *Collier Co.*—pond 0.3 mi. n. Everglades; pond 1.5 mi. n. Everglades; pond 1.8 mi. n. Everglades; pond 2.7 mi. n. Everglades; marsh nr. x-section U.S. 41 and Fla. 29; pond 5.2 mi. w.n.w. x-section U.S. 41 and Fla. 29; canal 8 mi. w.n.w. x-section U.S. 41 and Fla. 29. *Dade Co.*—canal 5.6 mi. n.w. Card Sound; Everglades National Park, West Lake; culvert 1 mi. n.e. West Lake; pond 7.7 mi. n.e. Flamingo; pond 3.4 mi. n.e. West Lake; pond 9 mi. n.n.e. Flamingo; pond 9.6 mi. n.n.e Flamingo; pond 10.6 mi. n.n.e. Flamingo; Nine-Mile Pond; Whiskey Creek; Paurotis Pond; Mahogany Hammock. *Monroe Co.*—Big Pine Key, quarry on main road 3 mi. n.w. U.S. 1; Cudjoe Key; Ramrod Key. *Lee Co.*—marsh 1.1 mi. e. x-section Fla. 865 and Fla. 867; creek 3.7 mi. e. x-section Fla. 865 and Fla. 867; Turners Slough, Sanibel Isl. (*MCZ*). *Sarasota Co.*—Warm Mineral Springs (*MCZ*).

ECOLOGY. *Heleobops docima* has been found in brackish marshes that had a Cl^- concentration of 1.14-19.98 ppt. The most extensive colonies and those containing the largest shells were found in waters with a Cl^- concentration of less than 5 ppt. At most stations where it was collected it was found in ponds in mangrove and sedge marshes. At such stations the water was generally less than 2 feet deep and subject to tidal fluctuations.

FIGURE 2

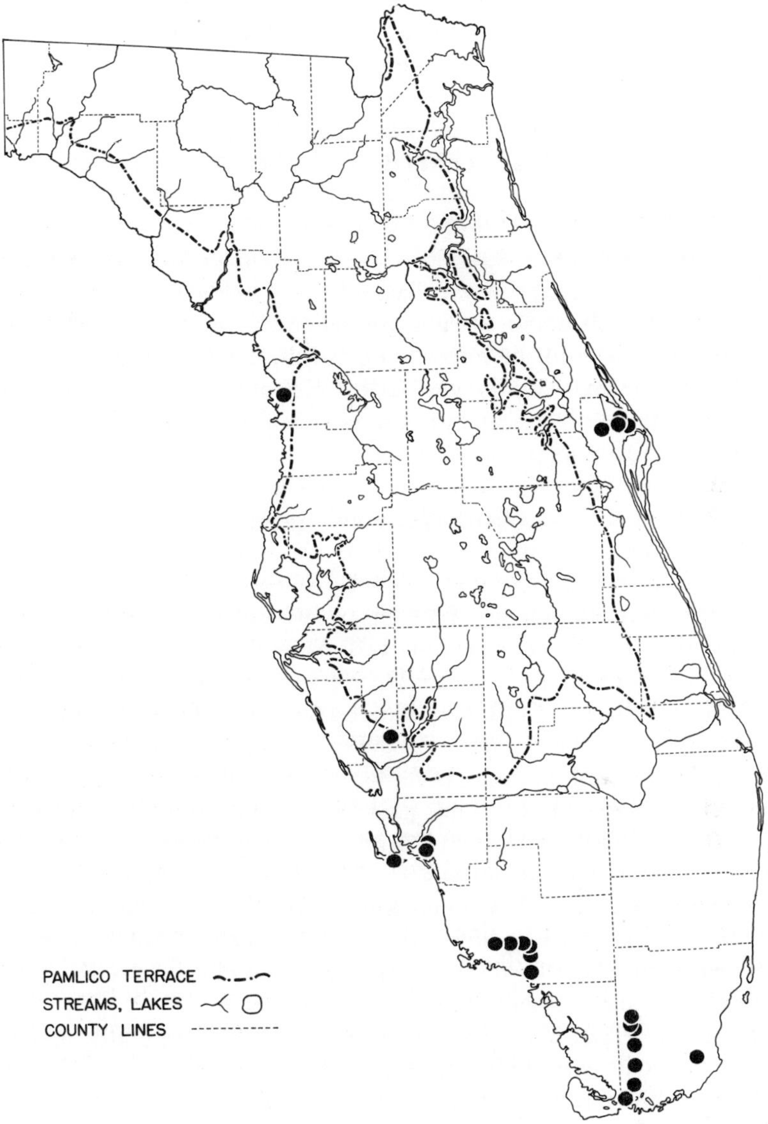

The geographic distribution of *Heleobops docima* new species on the Florida mainland.

Specimens were usually found on algal mats and rooted vegetation and were collected by shaking the plants in a pail of water. Occasional specimens were also found on the silty substrate that usually lies on the bottom of the brackish ponds in which they occur.

THE *ONOBOPS* TRIBE

This tribe is characterized by a simple elongate verge that lacks accessory appendages or glands. So far as is known the tribe contains only a single North American genus, *Onobops* (new genus). Some species of *Aphaostracon* (new genus) have a simple verge similar to that found in *Onobops,* but this is a secondary condition in the former genus, for those particular species apparently were derived from an ancestral stock that possessed papillae on the verge.

Onobops new genus

The shell is small, elongate-conical or cylindric-conical, narrow, 1.74-2.36 times as long as wide; variable in thickness, thin and transparent to thick and opaque. Spire narrow, with a flattened or rounded apex, elongate, 1.48-2.38 times the height of the aperture, weakly convex in outline. Umbilicus narrowly rimate or imperforate due to appressed columellar parietal margin of aperture. Periostracum light gray to light brown in color, dull. Sculpture consisting of numerous spiral threads that vary from subobsolete to coarse, about 10-25 threads between the sutures on the penultimate whorl. The threads may be interrupted by irregular raised incremental striations. 4.6-5.7 whorls; 2.00-2.45 whorls/mm length; nuclear whorl 0.10-0.15 mm in diameter; first one and one-half planular; remaining whorls rapidly descending to the aperture at an increasing rate; whorls weakly arched or nearly flat-sided. Suture deeply incised, tending to form a channel. Aperture ovate, simple in structure; at an oblique angle of about 30° to axis of shell. Peristome complete across parietal margin at maturity, adnate to the preceding whorl; outer lip thin and fragile or thick with an inner callus; parietal margin rather long and straightened.

Length, 1.70-2.70 mm; diameter, 0.90-1.35 mm; height of aperture 0.60-1.00 mm; width of aperture, 0.50-0.80 mm.

Operculum chitinous, very thin, fragile, transparent, colorless. Paucispiral, consisting of about 3-4 whorls, the outer of which expands more rapidly than do the earlier whorls. Outer surface sculptured with numerous fine incremental striations. Inner surface smooth.

Both species of *Onobops* are very much alike in general anatomical characteristics (*O. jacksoni*: Fig. 27, E). The foot is relatively short and spatulate with the anterior corners projecting laterally and the sole constricted at the base of the pedicel. The anterior margin is truncate and the posterior end is bluntly pointed. The tentacles are about as long as the snout and have about 10 concentric ciliary rings. The eyes are located at the outer base of the tentacles. The snout, sides of the body, nape, and tentacles are lightly pigmented with brown. The tip of the snout and the tips of the tentacles are white. The upper 4 whorls of the body are strongly mottled with brown along their upper surfaces. The coloration is variable and some specimens may have hardly any indication of a color pattern. The gill extends from the center of the mantle collar posteriorly and laterally to the pericardium, and consists of 20-28 lamellae. The lamellae are most strongly developed in the middle of the series and are slightly reduced anteriorly and more so posteriorly.

The female is oviparous. The verge of the male originates on the nape slightly to the right of the middorsal line at a distance behind the tentacles slightly greater than the length of the snout. When contracted the verge is coiled counterclockwise on the nape beneath the mantle. It is simple in structure, elongate, and dorsoventrally compressed; it lacks appendages and surface glands. The tip of the verge tapers to a fine point. The vas deferens is proportionally large, simple, and lacks accessory ducts.

The radula is a very small narrow ribbon, which in specimens with about 5 whorls reaches a length of about 0.4 mm. It is taenioglossate with 47-56 transverse rows of teeth. The central tooth is trapezoidal with a medium ventral projection, 5 ectocones on each side of the mesocone, and a single basocone on each side of the face. The lateral tooth is clavate with an elongate shaft; it has 2-3 entocones, a larger mesocone, and 4-5 ectocones. The inner marginal tooth has a moderately long flange along the inner margin of its shaft, and has about 18-20 cusps on the rasping margin. The outer marginal tooth has a relatively wide, short flange-like projection along the basal half of the shaft, and has about 20 minute cusps that are too small to be clearly distinguished under oil immersion.

Type species: *Onoba crassa* new species.

The name *Onobops* refers to the similarities of the shell to that of the marine rissoid *Onoba*.

Onobops is known to occur from the Chesapeake Bay region to the Florida peninsula, and probably has a wider distribution along the Atlantic and Gulf Coasts of eastern North America. It is characterized by a small elongate-conical or cylindric-conical shell with a long spire that has a relatively large rounded apex and a minute, planular nuclear whorl. The umbilicus is narrowly rimate or closed, and the suture is weakly impressed, but channeled. The sculpture consists of numerous fine spiral threads that are frequently interrupted and may be subobsolete. The aperture is simple and oblique. *Onobops* is anatomically distinct from other New World genera of the family Hydrobiidae by being oviparous and by having a simple, long, slender, compressed verge.

This genus does not show close affinities to any of the known American hydrobiid genera. The European genus *Hydrobia* is most similar to *Onobops*, but *Hydrobia* differs by having a smooth shell devoid of spiral sculpture. *Aphaostracon* is also similar to *Onobops* in the shape of the shell and the structure of the male reproductive system, but it differs by having a smooth shell with only incremental striations, and by having a rounded verge that typically has papillae along the right margin.

Onobops is known to include two species that are distinguished from each other as follows.

O. jacksoni (Bartsch): The shell is cylindric-conical, thin, fragile, and transparent. The spiral sculpture is usually strongly developed. The outer lip is thin and fragile and never has an internal deposit. The gill consists of about 20-24 lamellae, and the operculum has about 3 whorls.

O. crassa new species: The shell is elongate-conical, thick, and translucent or opaque. The spiral sculpture is much reduced and may be nearly obsolete. The outer lip is thick and usually has a heavy internal callus at maturity. The gill consists of about 26-28 lamellae and the operculum has about 4 whorls.

Onobops jacksoni (Bartsch)

Onoba jacksoni Bartsch, 1953; Naut., 67: 40-41; pl. 5, fig. 5.

SHELL (Fig. 58, D; Fig. 50, G, H). Small, elongate, cylindric-conical, terreted; very thin, fragile, transparent or translucent. Spire 1.63-2.38 times the height of aperture, slightly convex in outline; apex bluntly rounded. Umbilical chink very narrowly rimate or imperforate due to appressed columellar-parietal angle. Periostracum light gray or gray brown with a dull luster. Sculpture consisting of numerous rather coarse spiral threads that may occasionally be broken or raised on irregular incremental striations. Rarely is the sculpture reduced to the extent that it is difficult to discern, 4.6-5.7 whorls; 2.08-2.45 whorls/mm length; nuclear whorl 0.10-0.15 mm in diameter; first one and one-half whorls planular; remaining whorls descending to the aperture at an increasing rate; sides of whorls very slightly arched, nearly flat but with a weak shoulder along the suture. Suture deeply incised, forming a channel. Aperture ovate; upper corner bluntly rounded; interior hyaline; axis of aperture at about 20° to axis of shell; plane of aperture at about 30° to shell axis. Peristome usually complete across parietal margin in large specimens; outer lip thin, fragile, evenly curved from the upper corner to the columellar margin, very slightly sinuous in lateral profile; parietal margin thin, long, nearly straight.

Length, 1.70-2.90 mm; diameter, 0.90-1.40 mm; height of aperture, 0.60-1.07 mm; width of aperture, 0.50-0.80 mm.

OPERCULUM (Fig. 28, H). Small ovate, chitinous, very thin, fragile, transparent, colorless. Paucispiral, consisting of about 3 whorls. The first 2 are tightly coiled. The last whorl rapidly expands in size. Nucleus located at about one-third of the distance from the left margin to the right and at about one-fourth of the distance from the base to the apex. Outer surface with numerous very fine wavy incremental striations. Inner surface smooth.

VERGE (Fig. 37, I). In all important features the verge is similar to that of *O. crassa.*

RADULA. The ribbon contains 51-56 transverse rows of teeth. The central tooth has 5 ectocones and a single basocone on each side of the mesocone. The lateral tooth has 3 entocones, a single large mesocone, and 4-5 ectocones. The inner marginal tooth has 18-20 cusps. The outer marginal tooth has about 20 cusps.

TYPE LOCALITY. Town Point, on the Little Choptank River, near Dailsville, Dorchester Co., Maryland. Type: *USNM* 361662.

DISTRIBUTION (Fig. 3). Known only from brackish marshes along the gulf coast of Levy County, Florida, and the Chesapeake Bay region of Maryland.

LOCALITIES. Florida: *Levy Co.*—pond near the intersection of Fla. Hwy. 24 and Fla. Hwy. 347; marsh 3.2 mi. s.w. Yankeetown. Maryland. *Dorchester Co.*—Town Point, on the Little Choptank River (paratypes: *USNM* 361663). Anatomical material examined only from Levy Co., Florida.

ECOLOGY. This species was found in small, shallow brackish pools in sedge marshes. The pools contained thick mats of aquatic angiosperms and filamentous algae. Snails were collected by shaking plant mats in a pail of water.

Onobops crassa **new species**

SHELL (Fig. 58, E; Fig. 50, I, J). Small, elongate-conical, 1.74-2.12 times as long as wide, conspicuously thickened, subtranslucent or opaque. Spire very long, slender, conical with a blunt apex, 1.48-1.95 times the height of the aperture; slightly concave. Umbilicus narrowly rimate or imperforate. Periostracum varying from light gray to light brown, with a dull luster; edge of columellar lip frequently dark brown. Sculpture consisting of numerous very fine, barely distinct raised spiral threads which are frequently interrupted by rather coarse, irregular incremental striations, about 15-25 threads on penultimate whorl. In some specimens the spiral sculpture may be reduced to the extent that it is barely detectable and may persist only along the suture and around the umbilicus. 5.0-5.4 whorls; 2.00-2.27 whorls/mm; nuclear whorl 0.10-0.15 mm in diameter; first one and one-half whorls planular; remaining whorls descending to the aperture at an increasing rate; periphery of whorls weakly arched, nearly flattened. Suture deeply incised, almost channeled. Aperture ovate; upper corner thickened, bluntly angulate; interior livid white due to thickness of shell; axis of aperture at about 20° to axis of shell; plane of aperture at about 30° to shell axis. Peristome complete across parietal margin at maturity; outer lip thick and with a heavy internal callus, rounded from the upper corner to the columellar margin, very slightly sinuous in lateral profile; columellar margin appressed over umbilical

FIGURE 3

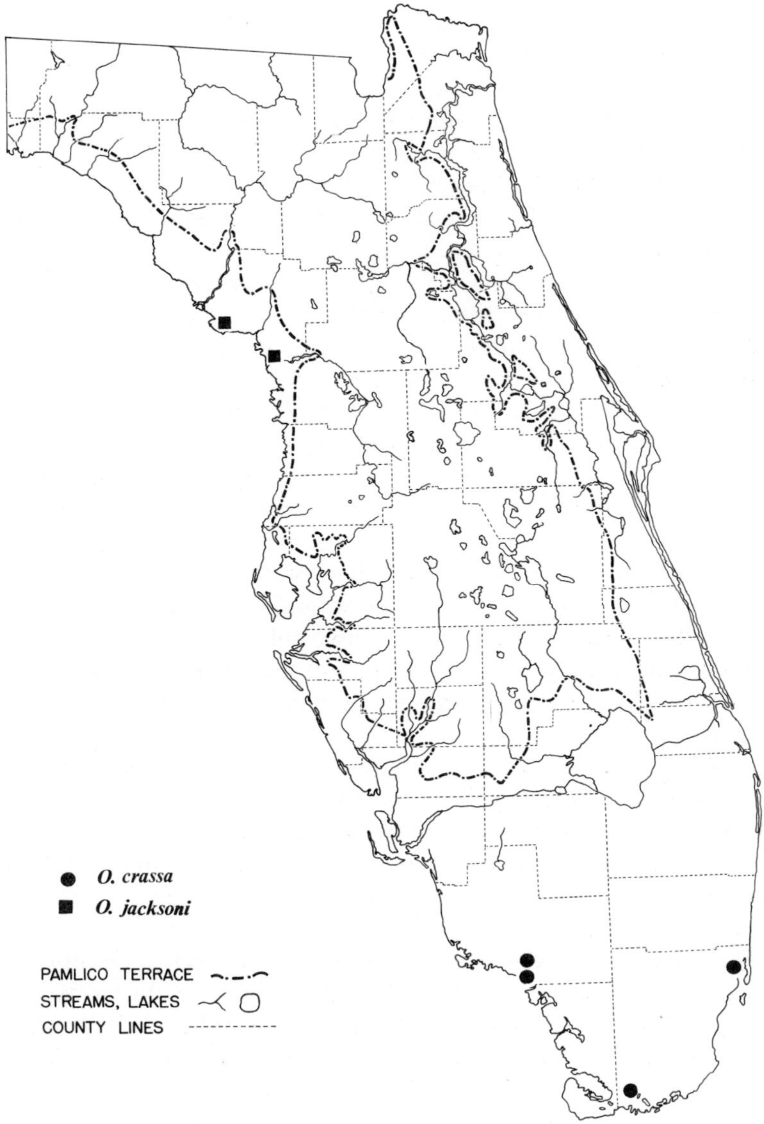

The geographic distribution of *Onobops jacksoni* (Bartsch) and *Onobops crassa* new species in Florida.

region leaving at most a very narrow chink; parietal margin relatively long and nearly straight.

Length, 2.20-2.70 mm; diameter, 1.10-1.35 mm; height of aperture, 0.80-1.00 mm; width of aperture, 0.65-0.80 mm.

Measurements of type: length, 2.40 mm; diameter, 0.96 mm; height of aperture, 0.91 mm; width of aperture, 0.72 mm; 5.3 whorls; nuclear whorl 0.12 mm in diameter.

This species has a characteristic appearance even though it is highly variable in relative width, the relative size of aperture, and relative number of whorls per length.

OPERCULUM (Fig. 28, G). Small, ovate, relatively narrow, chitinous, very thin, fragile, transparent, colorless. Paucispiral, consisting of about 4 whorls. The inner 3 whorls are tightly coiled and the outer whorl rapidly expands in size. Nucleus located in the lower left third of the operculum. Outer surface with numerous very fine incremental striations. Inner surface smooth.

VERGE (Fig. 37, H). Originating on the nape very slightly to the right of the middorsal line and at a distance behind the tentacles slightly greater than the length of the snout. It is simple in structure, long, tapered at the tip, and dorsoventrally compressed. The vas deferens is simple, relatively thick, and convoluted. When contracted the verge lies in a circular arc coiled counterclockwise beneath the mantle. When erect it extends anteriorly over the dorsal surface nearly to the tip of the snout and between the tentacles or off to the right side.

RADULA (Fig. 33, D). The ribbon contains 47-52 transverse rows of teeth. The central tooth has 5 ectocones and a single basocone on each side of the mesocone. The lateral tooth has 2-3 entocones, a single mesocone, and 4 ectocones. The inner marginal tooth and the outer marginal tooth both have about 20 minute cusps along the rasping margin.

TYPE LOCALITY. Mangrove marsh 1.9 miles north of Everglades, Collier County, Florida. Type: *UMMZ* 216537; collected April 9, 1963 by Fred G. Thompson. Paratypes: *UMMZ* 216538 (133), *USNM* 636947 (5), *ANSP* 301539 (5), *MCZ* 258005 (5), *CM* 65-12 (4), *UF*.

DISTRIBUTION (Fig. 3). Known only from coastal marshes in Dade and Collier Counties, Florida.

LOCALITIES. Florida: *Dade Co.*—North Miami Beach, Oleta Rd. and U.S. 1; pond near West Lake, Everglades National Park. *Col-*

lier Co.—pond 1.8 mi. n. Everglades; marsh at x-sec. U.S. 41 and Fla. Hwy. 29.

ECOLOGY. This species is confined to mangrove swamps, where it is found in quiet pools over a soft muck bottom. It is usually buried on the bottom in a diatomaceous ooze and is very seldom found crawling on aquatic vegetation. The type locality, where it was collected on two occasions, is the only station at which it was found to be common. It is usually associated with *Littoridinops monroensis* and *Pyrgophorus platyrachis*.

THE *HYDROBIA* TRIBE

This tribe contains four American genera which may be difficult to distinguish due to the high degree of variability that exists in the shells. They may be diagnosed as follows.

Pyrgophorus. Shell conical or cylindric-conical, thin or moderately thick, translucent or opaque, olivaceous to brown in color; spiral sculpture consisting of raised threads, the uppermost of which may bear a series of conical or triangular spines; peristome complete. Shell demonstrating slight sexual dimorphism. Females ovoviviparous, with about 50 small embryos in the uterus. Verge with 3-7 papillae along right margin; left margin usually with a small papilla near base, and a projection near end, which bears 1-4 papillae.

Hyalopyrgus new genus. Shell elongate-conical or cylindric-conical, thin, fragile, transparent, colorless to light olive gray; spiral sculpture consisting of fine incised striations, which are readily evident; peristome complete or incomplete. Shell demonstrating strong sexual dimorphism in size. Female ovoviviparous, with about 15 large embryos in uterus. Verge with a large papilla on right margin near base, and 1 or 2 smaller papillae on left margin near distal end.

Littoridinops. Shell conical, thin or moderately thick, translucent or opaque, olivaceous; spiral sculpture consisting of a few sparse, fine striations that are difficult to distinguish except with transmitted light; peristome usually incomplete, but may be complete and thick in one species. Shell not sexually dimorphic. Females oviparous. Verge with 7-50 papillae along right margin and 1-4 papillae along distal third of left margin; a series of papillae may also be present about base of verge.

Aphaostracon new genus. Shell small, conical or cylindric-conical, variable in thickness, transparent or opaque. Light to dark brown; sculpture consisting only of fine incremental striations, no spiral sculpture; peristome complete across parietal wall. Shell sexually dimorphic in size. Female oviparous. Verge with 0-6 papillae along right margin; no other papillae present. (The absence of papillae on the verge is a secondary feature due to the loss of such structures within the genus).

These four genera appear to be more closely related to each other because of the structure of their verges than to any of the other genera that inhabit Florida. Within this tribe they represent a diverse phyletic assemblage.

The European genus *Hydrobia* (type species *H. ventrosa* Montagu) differs from these genera in having a single papilla near the middle of the left margin of the verge. In other respects the verge is similar to that of *Aphaostracon* (Muus, 1963).

Pyrgophorus Ancey, 1888

Pyrgophorus Ancey, 1888, Bull. Soc. Mala. France, 5: 192. Type species by original designation: *Pyrgulopsis spinosa* Call and Pilsbry, 1886.

The shell is medium-sized, conical or cylindric-conical; thin to moderately thick; translucent or opaque, very seldom transparent. Spire elongate, conical, usually longer than aperture; straight-sided or convex. Umbilicus rimate or imperforate. Sculpture consisting of fine incremental striations; spiral sculpture may be absent, but usually present, consisting of spiral lirae that vary from fine threads to strong chords. Triangular or conical spines may be present on the uppermost spiral chord; the spines may be dorsoventrally flattened, conical, or laterally compressed. Spiral sculpture developed in both the periostracum and the shell. 5.0-6.5 whorls; embryonic whorl strongly protruding, 0.20-0.25 mm in diameter. Suture deeply impressed, regularly descending to the aperture; whorls strongly rounded, but usually not shouldered. Aperture ovate. Peristome complete at maturity; outer lip and basal lip thin, sharp; columellar margin thin, weakly reflected over umbilicus, tending to obscure or occlude it; parietal margin broadly in contact with preceding whorl, thin. Periostracum olivaceous brown to light brown, occasionally very light or colorless.

Operculum thin, transparent, chitinous; amber yellow, ovate.

Paucispiral, consisting of 2.0-2.5 rapidly expanding whorls; nucleus located about one-fourth of the distance from the columellar margin to the outer lip margin, and about one-third of the distance from the basal margin to the apex.

Radula typically taenioglossate. Central tooth with 5-6 aculeate ectocones on each side of the mesocone on the reflected margin, and with 2-3 basocones on or at the base of the lateral extensions; basal margin with a simple lobe-like median projection. Lateral tooth with 3-1-5 to 5-1-7 cusps on the reflection; body of lateral tooth squarish with a long slender shaft projecting ventrolaterad. Inner marginal tooth with 20-35 cusps on the reflection; shaft of tooth with a flange-like extension along basal half of inner margin. Outer marginal with 20-30 cusps on the reflection; shaft with a narrow flange-like extension along basal half of outer margin.

Operculum supported in part by fleshy triangular lobes that extend laterally beyond the margins of the foot. Basal half of left tentacle with 10-15 bands of cilia that encircle tentacle. Female ovoviviparous; young snails with about 1-2 whorls at time of birth. Males with a dorsoventrally compressed verge that originates on the right side of the nape, verge terminating with a slender conical penis; right margin of verge with 3-7 short papillae; left margin with or without a single small papilla near the base, and with a large lateral projection at or near the end, which bears 2-4 short, conical papillae; vas deferens simple. The papillae along the margins of the verge have several hollow chambers that are radially arranged, and separated by connective tissue and muscle fibers. The chambers are continuous with the sinuses of the penis.

Pyrgophorus is confined to the circum-Caribbean region. It is known from Texas south to Panama, Colombia, and Venezuela, and the West Indies, exclusive of the Bahama Islands. It also occurs in southern Florida.

This genus has a confused nomenclatorial history, for all of the species have been described under different generic names. Pilsbry (1891, 327) placed the species in *Potamopyrgus* Stimpson, 1865, because of the strong spiral sculpture on the shell. Ancey (1888, 187) placed the American species in *Pyrgulopsis* Call and Pilsbry, 1886, but erected the section *Pyrgophorus* for those members with spiral threads bearing spines. Pilsbry (1891, 327-28; 1911, 562) followed his earlier course, and relegated the name *Pyrgophorus* to the synonymy of *Potamopyrgus*. H. B. Baker (1930, 31) followed Pilsbry in associating the American species with *Potamopyrgus*, but recognized *Pyrgophorus* as a distinct subgenus. Morrison (1939, 87) separated *Pyrgophorus* from *Potamopyrgus*, but placed it in the synonymy of *Lyrodes* Doering, 1884, because in both groups the spiral sculpture is developed in the shell as well as in the periostracum. Taylor (1966, 192) recognized *Pyrgophorus* as distinct, but associated it with the neotropical genera related to *Heleobia.*

Doering's description of *Lyrodes* (1884, 463-64) indicates that its type species is not ovoviviparous, for mature embryos in the uterus would have been too conspicuous to have been overlooked with the detail that he used to describe it. Morrison (personal communication, 1962) has informed me that the verge of *Lyrodes* is similar to that of *Heleobia* Stimpson, 1865, in possessing large glandular tubercles along the margins. The glandular developments on the verge of *Lyrodes* relate it to *Heleobia* and other genera confined to South America. The hollow, multichambered papillae on the verge of *Pyrgophorus* relate this genus to *Littoridinops* and *Hyalopyrgus.*

Pyrgophorus platyrachis new species

SHELL (Fig. 59, A-C). Thin, translucent, olivaceous brown to light brown, nuclear whorls white. Elongate-conical; spire 1.1-2.1 times the length of the aperture, straight-sided. Umbilicus open, but narrowly rimate, very seldom closed. Sculpture variable; axial sculpture consisting of numerous fine, close incremental striations; spiral sculpture may be present or absent; when present it may consist of 1-11 spiral threads, which vary from fine to coarse. The spiral threads begin on the spire shortly below the embryonic whorls and are continuous to or near the aperture. The uppermost thread is most strongly developed, and is the only thread present in most cases. It may bear 12-15 dorsoventrally flattened spines per whorl on the body whorl or the penultimate whorls, or both. The spines are broader than long. Rarely, the spiral thread with spines may be so strong that it causes the upper surface of the whorl to be flattened. 5.1-6.5 whorls; nuclear whorl strongly protruding, 0.20-0.25 mm in diameter; whorls regularly increasing in size and regularly descending. Suture deeply impressed; whorls strongly rounded but not shouldered. Aperture ovate; hyaline inside; upper corner usually rounded-angulate; axis of aperture lying at about 20° to axis of shell; plane of aperture parallel or nearly so to shell axis. Peristome continuous at maturity; outer lip thin, dark-edged, straight in lateral profile; outer lip, basal lip, and columellar margin rounded; columellar margin thin, weakly reflected over umbilical perforation, but not occluding it; parietal margin very thin, hyaline, broadly in contact with preceding whorl, not extending forward beyond plane of aperture.

Measurements of specimens with complete apertures: length of shell, 2.7-5.3 mm; diameter of shell, 1.6-2.9 mm; length of aperture, 1.1-2.0 mm; width of aperture, 0.9-1.6 mm; length/diameter, 1.54-2.08; length/aperture, 2.13-3.06; whorls/length, 1.13-1.85.

As with most other species of *Pyrgophorus,* specimens completely lacking spiral sculpture, as well as specimens with well-developed spiral threads and spines, may occur in the same locality. Usually specimens with only a single strong spiral thread predominate in a series. In brackish marshes bordering the Everglades region some populations may have individuals that are devoid of spiral sculpture, and the suture may be very deeply impressed along the early whorls so that they are almost free from each other.

Specimens with strong spiral threads bearing spines tend to be more obese than those lacking such sculpture. Females tend to be more obese than males. These features of obesity appear quite striking upon casual examination, but they are not consistent.

Operculum (Fig. 29, B). Ovate, thin, transparent, amber-white; chitinous. Paucispiral, consisting of about 2.3 whorls, which rapidly expand in size. Nucleus located about one-fourth of the distance from the columellar margin to the outer lip margin, and about one-third of the distance from the base to the apex. Outer surface with numerous delicate incremental striations. Inner surface smooth.

Animal (Fig. 25, A, B). The foot is spatulate; narrowest at the anterior margin of the pedicel and widest at the posterior margin; truncate anteriorly, bluntly pointed posteriorly. Operculum supported by operculigerous lobes that extend laterally beyond the margins of the foot. Tentacles slender, about twice the length of the snout when extended; basal half of left tentacle with 10-15 ciliary bands (exaggerated in illustrations). The gill occurs in a long narrow series consisting of about 30-33 lamellae; it becomes widest shortly behind the mantle collar, and uniformly tapers in width posteriorly. Snout, nape, and sides of pedicel heavily pigmented with melanophores, which become more diffuse on the dorsal and lateral surfaces of the foot. Numerous fine silver and golden flecks may be scattered through the darker pigmentation, and become very concentrated and intense around the base of the tentacles and along the sides of the body. The tentacles are dark, as are the nape and the sides of the pedicel, but about 10-25 large yellow spots are distributed through their lengths.

The female contains about 50 minute embryos in the uterus. At the time of birth the embryos are about 0.3 mm wide and have 1.2 whorls. They escape from the egg capsules prior to birth and are active within the uterus.

VERGE (Fig. 39, I-K). The verge originates in a transverse position on the nape, behind the tentacles, so that its left basal margin lies on the median line, and the distal end projects to the right and forward when erect. When contracted the verge is recurved to the left beneath the mantle, so that its distal end lies over its base. It is dorsoventrally compressed, and of about equal width throughout its length. The right margin bears a series of 5-7 short, blunt papillae. The series begins at the base and may extend nearly to the end. A single small, blunt papillae is located on the left margin near the base. The distal end of the verge is bifurcate. The right side ends as a long slender penis through which the vas deferens passes. The left side ends as an enlarged lateral or terminal projection upon which are located 2-4 papillae. All of the papillae on the left and right margins have several radially arranged sinuses that are continuous with the hemocoel. The vas deferens is simple and tubular. It enters the verge near the right margin of the base, and extends through the penis, following the right margin throughout its length.

RADULA (Fig. 32, A). Twenty specimens examined. The teeth are arranged in 60-73 transverse rows. The central tooth is about 25μ wide and about one-half as high. The cusp formula is $\frac{5\text{-}1\text{-}5}{2\text{-}2}$. Occasionally an additional ectocone may be present on either or both sides. The lateral tooth has 3-1-5 cusps. Both marginal teeth have about 20 cusps.

TYPE. *UMMZ* 216485: canal along Florida highway 706, 7.4 miles w.n.w. of Jupiter, Palm Beach County, Florida; collected March 23, 1962 by Fred G. Thompson. Paratypes: *UMMZ* 216486 ($\pm$ 500); same data as the type.

DISTRIBUTION (Fig. 4). It is known only from Florida, where it

FIGURE 4

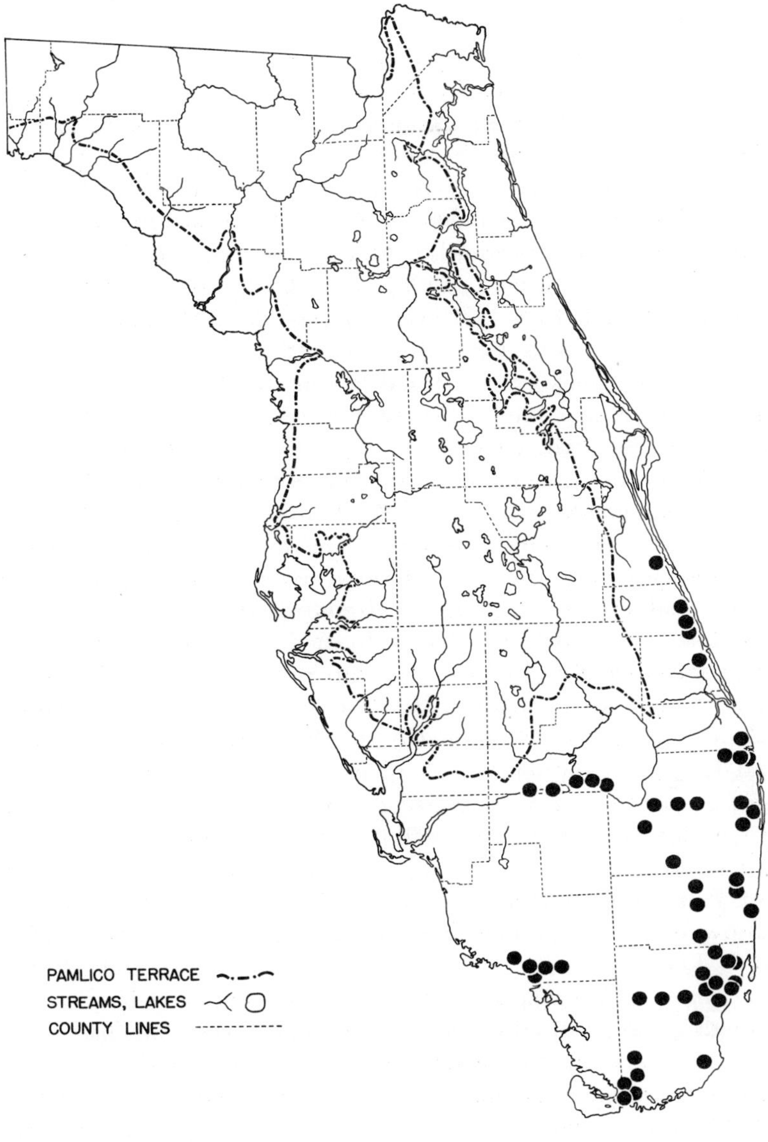

The geographic distribution of *Pyrgophorus platyrachis* new species.

has been found throughout the southern part of the peninsula from Lake Okeechobee south, and along the east coast as far north as southern Brevard County.

LOCALITIES. Florida: *Brevard Co.*—creek 0.3 mi. s. Grant. *Broward Co.*—canal 6.3 mi. n. Dade Co. line on U.S. 27; Andytown; canal 3.9 mi. n. Andytown; canal 4.1 mi. n. Margate; canal 5.4 mi. n. Margate; Ft. Lauderdale. *Collier Co.*—canal 8 mi. w.n.w. intersection U.S. 41 and Fla. 29; Turner R. 2 mi. e. Ochopee; canal 0.7 mi. e. Ochopee; canal at intersection of U.S. 41 and Fla. 29; canal 1.3 mi. w.n.w. intersection U.S. 41 and Fla. 29; canal 1.5 mi. n. Everglades; canal 1.8 mi. n. Everglades. *Dade Co.*—canal at Broward Co. line and N.W. 27 Ave.; Greynolds Park, North Miami Beach; numerous localities in Miami; canal 6 mi. w. Frog City; canal 9 mi. e. Coopertown; canal 0.5 mi. w. Coopertown; canal, Coral Gables; canal 4 mi. w. Kendall; canal 6 mi. n.w. Card Sound; Everglades National Park, West Lake; Nine-Mile Pond, Everglades National Park; pond 8.3 mi. n.e. Flamingo, Everglades National Park; pond 3.5 mi. s.w. West Lake, Everglades National Park; canal 1 mi. n.w. Pennsuco. *Glades Co.*—canal 4.5 mi. w. Clewiston; canal 6 mi. e.s.e. Moore Haven; canal 4.4 mi. e. La Belle; canal 11 mi. e. La Belle. *Hendry Co.*—Lake Okeechobee, Clewiston. *Indian River Co.*—canal 3.2 mi. s. Vero Beach; canal at n. edge Vero Beach; creek 3.2 mi. n. Vero Beach. *Lee Co.*—creek 1.1 mi. s.w. Ft. Myers. *Martin Co.*—Kitchen Creek, Jonathan Dickinson State Park. *Monroe Co.*—Cudjoe Key. *Palm Beach Co.*—canal 7.4 mi. w. Jupiter; canal 5.3 mi. w. Jupiter; pond 0.5 mi. s. Jupiter; canal 4.7 mi. w. Greenacre City; canal 14.4 mi. n.w. Broward Co. line along U.S. 27; canal 0.5 mi. e. Belle Glade; canal 6.1 mi. e. Belle Glade; canal 5.5 mi. w. Loxahatchee; canal 3.5 mi. n.w. West Palm Beach; canal 4 mi. w.s.w. Lake Worth; canal 4.4 mi. n. Broward Co. line along U.S. 27. *St. Lucie Co.*—creek 3 mi. s. Port St. Lucie.

ECOLOGY. This species is most commonly found in fresh water, although it has been collected in brackish marshes that had a Cl^- concentration of up to 6.5 ppt. It is found in a wide variety of habitats, but most commonly occurs in streams and canals that have a slight current and a soft, silty bottom. Some populations may be found in mangrove swamps. The snails usually inhabit places that have moderate or large amounts of organic material in the substrate. They are generally found on plants, but may also be found on any submerged object on the bottom. In many canals in south Florida, grass (*Panicum hemitomon*) and water hyacinths (*Eichhornia crassipes*) form floating mats along the shores or completely cover the waterway. In such situations *P. platyrachis* may be found in large numbers on the roots of the plants. In brackish water the snail is usually associated with *Chara* sp.

Turbidity of the water apparently has little effect upon the snail aside from governing the growth of vegetation upon which the snail lives.

RELATIONSHIP AND COMPARISONS. *Pyrgophorus platyrachis* is distinguished from *P. coronatus* (Pfeiffer) of Cuba by its thin, openly rimate shell, and by its flattened, broad, triangular spines, which are usually confined to 1.2 whorls or less. *Pyrgophorus coronatus* has a thick shell that is imperforate, and the spines are nearly conical, relatively elongate, and occur over a span of about 3 whorls.

Baker (1930, 32) described the verge of *P. parvulus* (Guilding) from Curaçao. Essentially it is like that of *P. platyrachis* except that it lacks a papilla near the base of the left margin. This difference may not be significant, for some populations of *P. spinosus* (Call and Pilsbry) contain individuals that lack the papilla although it is usually present among most members of a given colony.

The smooth form of *P. platyrachis* is apt to be confused with *Heleobops docima* and *Littoridinops monroensis,* as it is commonly found with these two species. *Pyrgophorus platyrachis* may be recognized by its more deeply impressed suture and its thin, complete peristome at maturity. Many populations of *L. monroensis* in brackish areas have a form that has a complete peristome, but in these instances the lip is conspicuously thickened.

Hyalopyrgus new genus

The shell is moderately large and sexually dimorphic in size; the male is about one-half or less the size of the female; elongate conical or cylindric-conical; thin, fragile; hyaline; olivaceous or glassy white. Umbilicate or imperforate. Spire equal to or greater than the length of the aperture, variable in outline. The sculpture consists of distinct, irregular incremental striations, and may include spiral striations or low varices. 4.5-7.5 whorls, which are separated by a deep suture; nuclear whorl 0.15-0.25 mm in diameter, slightly, or not at all protruding. Aperture oblong, ovate or nearly elliptical;

axis of aperture lying at about 30° to axis of shell; plane of aperture lying at about 10-20° to axis of shell. Peristome complete or incomplete across parietal margin; outer lip straight in lateral profile, very thin and fragile, continuously rounded with columellar margin. The columellar margin is weakly reflected, and may or may not close umbilical perforation.

Operculum ovate, thin, membranous, colorless, paucispiral, consisting of 2.5-3.5 whorls; nucleus located at about one-third of the distance from the columellar margin to the outer lip margin, and at about one-fourth of the distance from the base to the apex. Outer surface sculptured with fine, irregular, incremental striations.

The radula is typically taenioglossate, with 36-41 transverse rows. The central tooth is about twice as wide as long, and has a basolateral extension on each side and a midventral bulge; the reflection has 4-5 ectocones on each side of a large aculeate mesocone. There are 2-3 basocones located on the basolateral extensions. The upper basocone on each side is large and distinct; the lower basocones are located on a reinforced crest on the basolateral extensions, and may not be distinguishable except in lateral view. The lateral tooth is clavate, with 2-1-4 to 4-1-4 cusps. The inner marginal tooth has 11-20 cusps on the reflection. The outer marginal tooth has about 20-25 cusps on its reflection.

The foot is spatulate in shape, truncate anteriorly, bluntly pointed posteriorly and constricted at the base of the pedicel. The tentacles are slender, about 1.5-2.0 times the length of the snout when extended. The gill lies in a long series beginning at the collar and extending posteriorly to the pericardium. The series contains 21-29 lamellae which are largest near the collar, and taper progressively posteriorly. The gill of the male occupies the center of the mantle; in the female it is crowded to the left side to accommodate the enlarged uterus. The female is ovoviviparous and the uterus may contain about 10-15 embryos. The embryos have 2 whorls at the time of birth, and may be seen to move about within the uterus.

The verge is long, slender, and slightly tapered. It originates on the nape slightly to the right side of the midline. On its right margin it bears a large papilla at or near its base. On its left margin it bears 1 or 2 smaller papillae. In addition, a small lateral protuberance may be present on the left side of the apex. The end is blunt and terminates with a very small, slender, fleshy, pointed papilla. The papilla along the left margin contains several sinuses that are continuous with the hemocoel passages of the verge. The vas deferens is simple and terminates below the apical papilla.

Type species: *Bythinella aequicostata* Pilsbry, 1889.

Pilsbry (1934, 15) described the genus *Calipyrgula* from the Pliocene of California. In features of the shell *Hyalopyrgus* and *Calipyrgula* are not separable. Rather than considering it a relict of a previously much wider distribution, I suspect that *Hyalopyrgus* originated in Florida, and has no direct relationship to *Calipyrgula*. This suspicion is based upon general faunal dissimilarities between the two regions. Since only shells are preserved as fossils, almost any fossil hydrobiid is an enigma in its relationship to living forms, and to speculate on these relationships is futile.

Hyalopyrgus contains two species, *H. aequicostatus* and *H. brevissimus*, both of which are confined to peninsular Florida. Although they are quite different in appearance, they are considered congeneric. The shells have the same basic structure, though differing greatly in shape. They are thin, hyaline, similarly sculptured with incremental striations, and the nuclear whorls are nearly identical. The opercula differ only in minor details of outline and number of whorls. The radulae are very similar, differing consistently only in the number of basocones on the central tooth. The sexes are dimorphic in size. The females are ovoviviparous. The embryos of both species have the same number of whorls at birth and are about the same size. The verges differ only in the number and distribution of papillae. The differences that exist between the two species indicate that they are remotely related within the genus, but do not warrant separation at that level. The two species are distinguished as follows.

H. aequicostatus (Pilsbry) has an elongate-conical shell with 4.7-7.5 whorls, its peristome is in broad contact with the preceding whorl, and is usually incomplete, it is rimate or imperforate, its radula has 2 basocones on each side of the central tooth, and its verge has 2 papillae and an apical protrusion on the left margin.

H. brevissimus (Pilsbry) has a cylindric-conical shell with 4.5-5.1 whorls, its peristome usually is only in slight contact with or free from the preceding whorl, and is complete at maturity, its umbilicus is open, its radula has 3 basocones on each side of the central tooth, and its verge has a single papilla on the left margin.

The name *Hyalopyrgus* is derived from the Greek ὑαλόεις, meaning glassy, and πύργος, meaning tower, alluding to the thin, hyaline, elongate shell of the snail. The name is of the masculine gender.

Hyalopyrgus aequicostatus (Pilsbry)

Bythinella aequicostata Pilsbry, 1889, Proc. Acad. Nat. Sci. Phila.: 86-87; pl. 3, fig. 16.

Paludestrina aequicostata (Pilsbry), Pilsbry, 1899, Naut., 13: 22.

Fontigens oxybelis Pilsbry, 1950, Naut., 64: 37; pl. 3, figs. 1.

SHELL (Fig. 49, A-E; Fig. 59, D-F). FEMALE: Elongate, conical, narrow; thin, transparent; shiny, light olivaceous gray or brown to hyaline white; occasionally opaque white due to erosion of the shell. Imperforate or very narrowly rimate behind the reflected columellar margin. Spire long and conical, nearly straight-sided, about 1.5-2.7 times the length of the aperture. Sculpture consisting of fine incremental striations; superimposed upon these may be distinct spiral striations, or as many as 20 low broad varices per whorl; rarely do spiral striations and varices occur on the same shell. 4.7-7.5 whorls; first whorl protruding, 0.15-0.25 mm in diameter; remaining whorls regularly increasing in size; suture deeply impressed, regularly descending to the aperture. Aperture narrowly ovate or almost elliptical; longitudinal axis of aperture lying at about 30° to axis of shell; plane of aperture lying at about 20° to shell axis; interior of shell and aperture hyaline. Peristome incomplete, except in some very large specimens. Outer lip and basal lip continuously rounded with columellar margin; columellar margin reflected so that it obscures umbilicus, parallel to axis of aperture; parietal callus very thin, glassy.

Measurements of female shells: length of shell, 3.8-5.9 mm; diameter of shell, 1.65-2.6 mm; length of aperture, 1.2-1.9 mm; width of aperture, 0.9-1.5 mm; length/diameter, 1.82-2.68; length/aperture, 2.44-3.69; whorls/length, 1.02-1.53.

MALE: In most respects the shell of the male is like that of the female, except that it does not reach nearly as large a size (Tables 2 and 3). Generally the male is about one-half the length of the female, or less. It may usually be distinguished from a female shell of the same size by its more broadly flaring last whorl, and complete peristome. These differences are not consistent. In some populations all of the males may be recognized by these differences. In other populations none of the males may be distinguished by the shell.

TABLE 2.—MEASUREMENTS AND RATIOS OF FEMALE SHELLS OF *Hyalopyrgus aequicostatus* FROM SEVEN POPULATIONS. MEASUREMENTS ARE BASED UPON TEN SPECIMENS SELECTED FROM EACH SAMPLE. MEASUREMENTS ARE EXPRESSED IN MILLIMETERS

Locality	Length	Diameter	Aperture length	Whorls	Length/Diameter	Length/Aperture
Sanlando Springs, Seminole Co.	4.2 - 4.8	2.1 - 2.4	1.5 - 1.8	4.7 - 5.3	1.82 - 2.14	2.54 - 2.83
Blue Springs, Volusia Co.	4.4 - 5.0	2.2 - 2.6	1.6 - 1.9	5.5 - 6.1	1.83 - 2.23	2.44 - 3.06
Lake Tarpon, Pinellas Co.	3.8 - 4.5	1.9 - 2.4	1.4 - 1.8	4.7 - 5.6	1.87 - 2.07	2.50 - 2.71
Lake Panasoffkee, Sumter Co.	4.2 - 5.3	1.9 - 2.2	1.4 - 1.6	6.0 - 6.4	2.09 - 2.44	2.75 - 3.31
Creek 7.4 mi. s.e. Geneva, Seminole Co.	3.8 - 4.8	1.7 - 2.0	1.2 - 1.6	5.8 - 6.3	2.20 - 2.66	2.62 - 3.43
Lake Okeechobee, Martin Co.	4.9 - 5.3	2.0 - 2.4	1.5 - 1.8	6.3 - 7.0	2.18 - 2.68	2.83 - 3.38
Lake Apopka, Lake Co.	4.4 - 5.9	1.8 - 2.2	1.3 - 1.6	6.2 - 7.5	2.31 - 2.68	3.00 - 3.69

TABLE 3.—Measurements and Ratios of Male Shells of *Hyalopyrgus aequicostatus* from Six Populations. Measurements are Based Upon Ten Specimens Selected from Each Sample, and are Accurate to 0.05 Millimeters

Locality	Length	Diameter	Aperture length	Whorls	Length / Diameter	Length / Aperture
Blue Springs, Volusia Co.	2.00 - 2.25	1.15 - 1.55	0.95 - 1.15	4.0 - 4.3	1.45 - 1.74	1.96 - 2.27
Sanlando Springs, Seminole Co.	2.00 - 2.75	1.30 - 1.75	0.95 - 1.20	3.4 - 4.2	1.44 - 1.65	2.06 - 2.29
Lake Tarpon, Pinellas Co.	2.60 - 3.10	1.50 - 1.80	1.10 - 1.35	4.3 - 5.4	1.72 - 1.93	2.17 - 2.50
Lake Panasoffkee, Sumter Co.	2.60 - 2.80	1.20 - 1.50	1.00 - 1.20	4.8 - 5.0	1.87 - 2.16	2.33 - 2.70
Lake Okeechobee, Martin Co.	2.60 - 3.60	1.30 - 1.50	0.90 - 1.20	4.5 - 6.0	1.90 - 2.57	2.27 - 3.44
Lake Apopka, Lake Co.	2.50 - 2.95	1.20 - 1.45	1.00 - 1.15	4.9 - 5.3	2.03 - 2.20	2.45 - 2.70

Measurements of male shells: length, 2.00-3.60 mm; diameter, 1.20-1.80 mm; length of aperture, 0.90-1.35 mm; width of aperture, 0.70-1.00 mm; 3.4-6.0 whorls; length/diameter, 1.44-2.57; length/aperture, 1.96-3.44; whorls/length, 1.51-2.08.

Diversity of shell characters is greatly influenced by ecological conditions. Specimens from springs are usually more robust, have a larger aperture and have fewer whorls than specimens from lotic habitats such as lakes and sluggish rivers (Tables 2, 3). *Fontigens oxybelis* Pilsbry is an ecotype of *H. aequicostatus,* reflecting the spring form of the species. The development of spiral striations is most intense in snails from sluggish streams with a soft muck bottom. Varices usually occur in snails from sand-bottomed lakes. Snails from springs very seldom show evidence of spiral striations or varices. In most populations, some specimens lack either spiral sculpture or varices. Color is also influenced by the station occupied by the snail. Animals from springs and sluggish, soft-bottomed streams and lakes are light olivaceous gray or brown. Animals that occupy sand-bottomed lakes usually have a hyaline white shell.

OPERCULUM (Fig. 29, F). Apex pointed; parietal margin very slightly concave. Transparent. Consists of 2.5 rapidly expanding whorls. Outer surface finely, irregularly striate.

ANIMAL (Fig. 25, C, D). The tentacles are slender. The snout, pedicel, and foot may be unpigmented or they may be variously shaded with melanophores. The muzzle is white. The gill lies in a long series, which begins at the mantle collar. They are most developed immediately behind the collar, and taper in size posteriorly. There are about 22-26 lamellae in the series. In the female the gill is much narrower than in the male, and is more attenuate posteriorly.

The uterus of the female may contain about 15 embryos. Sexual maturity of the female is reached by the time they have acquired 4 whorls.

VERGE (Fig. 39, A-F). When extended the verge is about 1.5-2.0 mm long, slender, and is of nearly equal width throughout its length, tapering only slightly toward the end. It is slightly compressed dorsoventrally. The right side of the base is slightly swollen, and bears a large thumb-like papilla that is moderately shaded with melanophores. The end of the verge is blunt, and has a small conical papilla on its center, and a blunt lateral projection on the left corner. The left side of the verge on its distal third bears two medium-sized conical papillae. The vas deferens is relatively large in diameter and passes along the right margin of the verge.

RADULA (Fig. 32, B). Twenty specimens examined. Ribbon containing 36-41 transverse rows. Central tooth about 22μ wide. The cusp formula is constant, $\frac{4\text{-}1\text{-}4}{2\text{-}2}$; occasionally the lower basocone may be located over the thickened area of the lateral process, so that the cusp is not easily distinguished. Lateral tooth about 50μ long, and usually with 3-1-4 cusps. In some specimens only 2 entocones may be present, or a fourth ectocone may be found. Inner marginal tooth with 11-14 distinct cusps. Outer marginal tooth with about 25 cusps, which are too small to be clearly distinguished.

TYPE LOCALITY. The type locality for *Bythinella aequicostatus* is Sumter County, Florida. The type lot probably came from Lake Panasoffkee. Lectotype: *ANSP* 27985. The type locality for *Fontigens oxybelis* is Silver Springs Run (Silver Glen Springs), Marion Co., Florida. Holotype: *ANSP* 186751.

DISTRIBUTION (Fig. 5). This species is confined to the Florida peninsula. It is known from a zone across the upper half of the peninsula, which includes the lower half of the St. Johns River system, and the Withlacoochee River system south to Tampa Bay and the Orlando area. It is also abundant farther south in Lake Okeechobee. It may occur in some of the lakes in Polk and Osceola Counties, but numerous collections from these two counties have failed to reveal its presence.

LOCALITIES. Florida: *Brevard Co.*—7 mi. w. Mims. *Citrus Co.*—Crystal Springs. *Hernando Co.*—Weeki Wachee Springs. *Lake Co.*—Lake Apopka, Montverde; Lake Dora, 1 mi. s. Dora; Lake Dora, 1 mi. s. Tavares; Lake Harris at Fla. 19; Lake Eustis, Eustis; St. Johns R., Astor; many locations in Lake George (*MCZ*). *Marion Co.*—Silver Springs; Rainbow Spring Run, 2.8 mi. n. Dunnellon. *Martin Co.*—Lake Okeechobee, 3.7 mi. n. Port Myaca; Lake Okeechobee, 4.6 mi. n. Port Myaca. *Okeechobee Co.*—Lake Okeechobee, 3.9 mi. s. Okeechobee; Lake Okeechobee, 4.5 mi. s.e. Okeechobee. *Orange Co.*—Lake Apopka, 4 mi. s.w. Apopka; Lake Apopka, Winter Garden; Wekiwa Springs; St. Johns R. at Fla. 50; St. Johns R., 4 mi. e. Christmas; St. Johns R., 9 mi. w. Cocoa. *Pinellas Co.*—various localities in Lake Tarpon. *Putnam Co.*—Welaka Springs, 1 mi. n. Welaka. *Seminole Co.*—creek 7.4 mi. s.e. Geneva; St. Johns R., 5.4 mi. s.e. Geneva; Lake Jessup, 2 mi. n. Oviedo; north end Lake Jessup at Fla. 46; sulfur springs, Sanlando Springs. *Sumter Co.*—

FIGURE 5

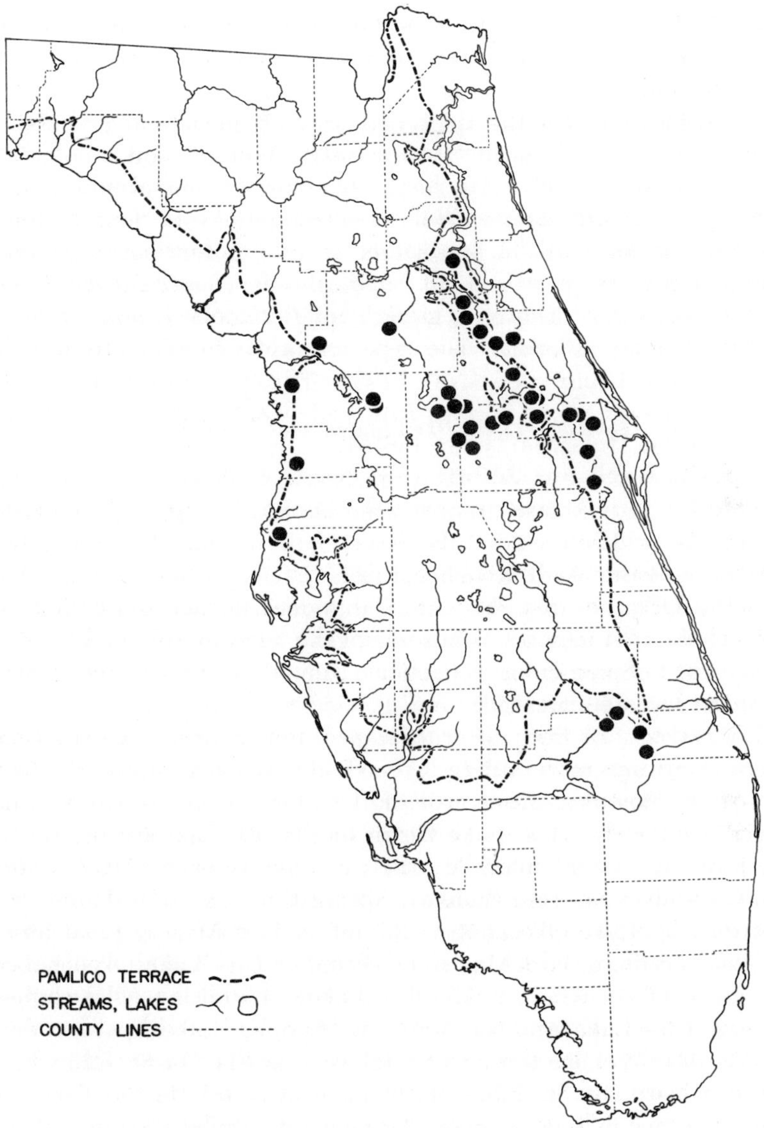

The geographic distribution of *Hyalopyrgus aequicostatus* (Pilsbry) in Florida.

Lake Panasoffkee, 3.7 mi. n.n.w. Panasoffkee; Lake Panasoffkee, 3 mi. n. Panasoffkee. *Volusia Co.*—creek 3.3 mi. w. Brevard Co. line on Fla. 46; St. Johns R., 3 mi. w. Orange City; Blue Springs, 3 mi. w. Orange City; St. Johns R., 4 mi. e. Sanford; Ponce de Leon Springs; Lake Woodruff.

ECOLOGY. This snail is apt to be found in most aquatic habitats within its range, except flatland marshes, sloughs, and small landlocked ponds or lakes. It has been collected in abundance in springs, lakes, rivers, and small streams, but shows no preference for any of these. The species is usually a bottom inhabitant, where it may be found on silt, sand, gravel, or large rocks. Occasionally it may be found on green plants or roots of floating vegetation. Water conditions may vary from clear to very turbid.

Hyalopyrgus brevissimus (Pilsbry)

Bythinella brevissima Pilsbry, 1890, Naut., 4: 64.
Paludestrina brevissima (Pilsbry) Walker, 1918, Misc. Pub. Museum Zool. Univ. Mich., (6): 137.
Amnicola harperi Dall, 1910, Naut., 24: 2.

SHELL (Fig. 49, F-I; Fig. 58, C). FEMALE: shell cylindric-conical; thin; transparent; olivaceous gray or grayish white; shiny. Umbilicate. Spire cylindric-conical, convex in outline, particularly near the top, about 1.0-1.5 times the length of the aperture. Sculpture consisting of fine, dense incremental striations; superimposed upon these are more widely spaced and weaker spiral striations, which are most prominent near the suture in immature shells (fewer than 4 whorls). 4.5-5.1 whorls in large shells; nuclear whorl protruding slightly, about 0.20 mm in diameter; remaining whorls rounded, globose, occasionally shouldered. Suture deeply impressed. Aperture oblong to ovate; oblique, its axis lying at about 30° to the axis of the shell; plane of aperture in lateral profile lying at about 10° to shell axis due to forward extension of the upper corner of the aperture. Interior of aperture and shell hyaline. Peristome complete; aperture free from or adnate to preceding whorl; when adnate, the parietal margin is short and thin; columellar margin parallel to axis of aperture, slightly reflected but not obscuring umbilicus; outer lip and basal lip thin.

Measurements of female shells; length of shell, 3.5-4.4 mm; diameter of shell, 2.2-2.4 mm; length of aperture, 1.6-1.8 mm; width of aperture, 1.2-1.4 mm; length/diameter, 1.52-1.83; length/aperture, 2.06-2.60; whorls/length, 1.16-1.32.

The description and measurements apply to large specimens. Females are sexually mature by the time they reach a length of 1.7 mm and have 3.2 whorls.

MALE: The maximum size of the male is minute compared to that of the female. The shells of both sexes are very similar in general structure, and mature males are very difficult to distinguish from immature females of the same size. The male shell is very short and conical, consisting of about 3.0-3.3 whorls. The umbilicus is broadly perforate. The spire is about equal to or less than the length of the aperture. The peristome is complete, and is free from or only slightly attached to the preceding whorl. The aperture is oval in shape, and in lateral profile its plane lies at about 20° to the axis of the shell.

Measurements of male shells: length of shell, 1.46-1.73 mm; diameter of shell, 1.35-1.48 mm; length of aperture, 0.75-0.96 mm; width of aperture, 0.70-0.79 mm; length/diameter, 1.07-1.15; length/aperture, 1.80-2.04.

OPERCULUM (Fig. 29, G). Translucent. Paucispiral, consisting of about three and one-quarter whorls. Nucleus very small. Outer surface with irregular, but usually fine delicate incremental striations.

ANIMAL. The snout and tentacles are moderately long. The snout is unpigmented, or only slightly gray. The nape and the sides of the pedicel are moderately pigmented with melanophores. The coloration is continuous on the sides of the tentacles to their tips. The gill series contains 21-29 lamellae, which are largest just behind the collar and become progressively smaller posteriorly.

The female is ovoviviparous, and large specimens may contain about 10 embryos in the uterus. The embryos have 2 whorls at the time they are released. Females become sexually mature by the time they have acquired 3 whorls.

VERGE (Fig. 39, G, H). The verge is about 1.2 mm long and is relatively slender as compared with other species within the genus. The base is slightly swollen immediately above the point of origin, and the verge gradually tapers to its blunt tip, which bears a small pointed papilla. A short blunt papilla is on the right side of the verge at about one-third of the distance from the base to the tip. Another papilla, about one-half as large, lies on the opposite side at about one-fifth of the distance from the tip to the base. The vas deferens is simple and tubular, and passes through the center of the verge.

RADULA (Fig. 32, C). Ten specimens examined. Radula with 37-41 transverse rows in females and about 30 rows in males. Central tooth about 25μ wide and 15μ high. Its cusp formula is $\frac{4\text{-}1\text{-}4}{3\text{-}3}$ or $\frac{5\text{-}1\text{-}5}{3\text{-}3}$. The two lowest cusps of the basocones are located on crests lying on the face of the transverse processes, and are difficult to see, particularly if the cusps have been worn. The upper cusp of this series is much more conspicuous, and upon superficial examination it may appear to be the only cusp present among the basocones. Lateral tooth with 4-1-4 cusps. Mesocone much larger than other cusps. Entocones about as long as, but more slender than ectocones. Two marginal teeth each with about 20 cusps.

TYPE LOCALITY. Pilsbry (1890, 64) stated that the type of *Bythinella brevissima* was collected by John H. Campbell, Esq., in the Haulover Canal at the head of the Indian River, where it was associated with *Hyalopyrgus aequicostatus, Notogillia wetherbyi, Goniobasis papillosa* (= *Goniobasis floridensis*), and *Helisoma scalare,* among other species. This locality undoubtedly is in error, since none of these five species occurs east of the St. Johns River system.

The type lot of *Amnicola harperi* came from the Pleistocene marl of Lake Panasoffkee, Sumter County, Florida (Dall, 1910: 2).

The type of *Bythinella brevissima* is *ANSP* 62418. It is a large female shell and is subfossil. The lectotype of *Amnicola harperi,* by present designation, is *USNM* 211011. It is a mature male. This is clear because of the characteristic sexual dimorphism in the shells, and because of the continuous distribution from the late Pleistocene marl into the present in Lake Panasoffkee.

DISTRIBUTION (Fig. 6). This species is known only from a small area in the central region of the Florida peninsula, where it has been collected in Orange, Seminole, Sumter, and Volusia Counties.

LOCALITIES. Florida: *Orange Co.*—Lake Apopka, 4 mi. s.w. Apopka; Lake Apopka at Winter Garden. *Seminole Co.*—St. Johns R., 5.4 mi. s.e. Geneva; Lake Jessup, 2 mi. n. Oviedo. *Sumter Co.*—Lake Panasoffkee, 3.7 mi. n.n.w. Panasoffkee; Lake Panasoffkee, 3 mi. n. Panasoffkee. *Volusia Co.*—St. Johns R., 3 mi. w. Orange City; Lake Woodruff, at De Leon Springs.

ECOLOGY. This species is sparse and sporadic in distribution. The

FIGURE 6

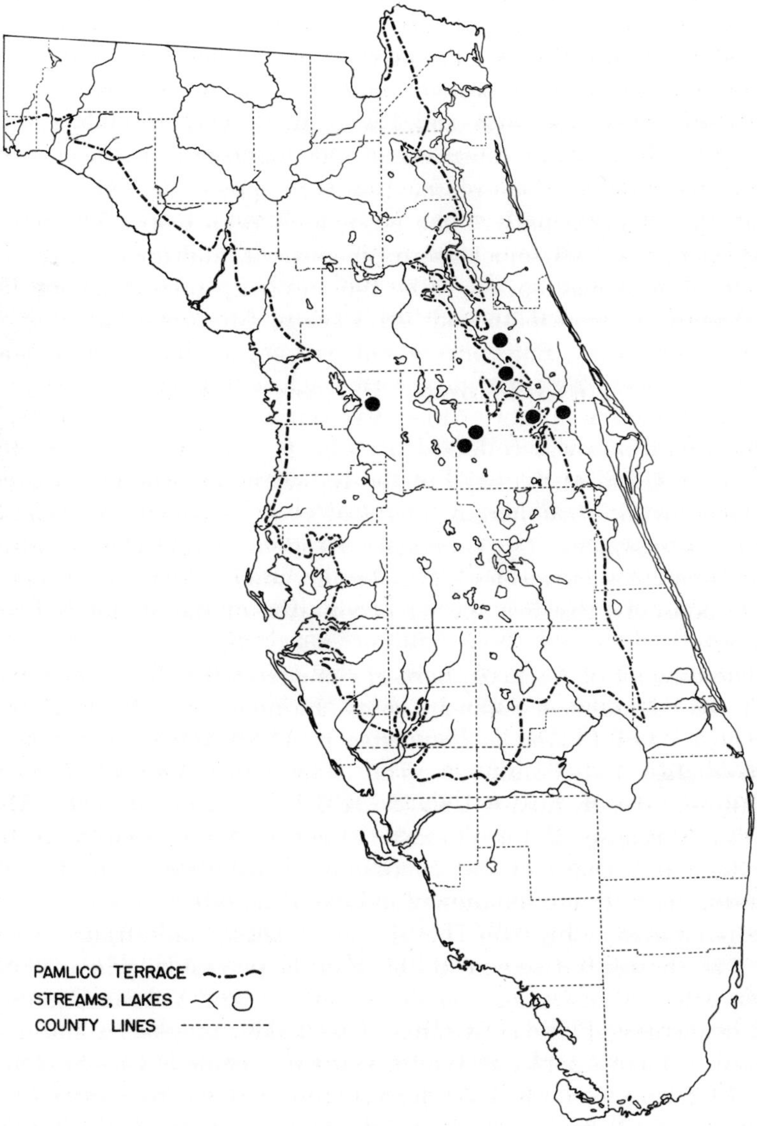

The geographic distribution of *Hyalopyrgus brevissimus* (Pilsbry).

only locality at which it was collected in abundance was in Lake Panasoffkee, 3.7 miles north of Panasoffkee, Sumter County, where it was very common on soft silty sand in about 5 feet of water. At all other localities it was found in very soft ooze that contained large amounts of dead organic material. Specimens were usually found in turbid water over 4 feet deep, and devoid of green vegetation, including algae; however, at Lake Woodruff a single specimen was found in shallow clear water in a patch of *Elodea*.

Males are much scarcer than females. From a total of about 300 specimens collected, only eight were males. This may be due to the small size of the males, for they could easily have been overlooked, or could have passed through the mesh of the collecting sieves.

Littoridinops Pilsbry, 1952

Littoridinops Pilsbry, 1952, Naut., 66: 51.

Type species by original designation: *Amnicola tenuipes* Couper.

Shell simple, medium-sized, conical or cylindric-conical; thin to thick, transparent, translucent, or opaque; periostracum brown, olivaceous, or white, usually with an oily luster. Umbilicate, rimate, or imperforate. Spire conical, or cylindric-conical, 0.75-2.00 times the length of the aperture. Sculpture consisting of fine incremental striations crossed by even finer, sparse spiral striations; sculpture may be very faint or obsolete. 4.5-6.8 whorls; nuclear whorl protruding, 0.10-0.23 mm in diameter; remaining whorls flat-sided or only slightly arched. Suture weakly impressed. Aperture ovoid or elliptical; axis of aperture lying at about 20° to axis of shell; plane of aperture lying at about 10-20° to shell axis. Peristome usually discontinuous across parietal wall; when continuous it may be adnate to or free from preceding whorl; outer lip simple, thin, sharp, occasionally with a callused deposit within; outer lip straight or nearly so in lateral profile; columellar lip thin, weakly reflected.

The operculum is ovate, thin, transparent, colorless or grayish white, chitinous, and paucispiral, consisting of about 2 rapidly expanding whorls. Its nucleus is large and lies at about one-third of the distance from the columellar margin to the outer lip margin, and at about one-fourth of the distance from the base to the apex. Its outer surface is sculptured with numerous fine incremental striations. Its inner surface is smooth.

The radula is typically taenioglossate, with basocones on the face of the central tooth. The teeth occur in 41-56 transverse rows. The central tooth is 25-42μ wide and about one-half as high. Its basal margin bears a large median tongue-like bulge, and each side has a thickened basolateral extension. The cusp formula varies from $\frac{4\text{-}1\text{-}4}{1\text{-}1}$ to $\frac{6\text{-}1\text{-}6}{3\text{-}3}$, usually $\frac{4\text{-}1\text{-}4}{2\text{-}2}$. The mesocone is elongate and dagger-shaped. The ectocones and the basocones are considerably smaller. The lateral tooth is hatchet-shaped, with a long lateral shaft. The basal margin is reflected partially over the lower half of the face. The cusp formula is 2-1-3 to 3-1-4, usually 2-1-4. The mesocone is about twice the size of the other cusps. The inner marginal has 14-20 cusps along the reflection. The shaft bears a long, thin, flange-like lateral extension. The outer marginal tooth has 15-23 cusps along the reflection. Its shaft has a conspicuous lateral flange-like extension along the basal half.

The foot is elongate and spatulate. It is widest at the posterior margin of the pedicel and is constricted near the middle of the pedicel. Its posterior end is bluntly pointed, and its anterior end is truncate with the two anterior corners extending as wing-like projections. The operculigerous lobes extend laterally over the sides of the foot as a small triangular flap on each side. The snout is blunt and trunk-shaped in live animals, capable of considerable extension beyond the anterior margin of the foot. The tentacles are very long and slender, twice the length of the snout in active animals, and each bears a diffuse band of pigmentation near the tip. The gill lies in a long series between the mantle collar and the pericardium. There are 31-40 lamellae in the series. The filaments are about equal in size, except near the pericardium and the mantle collar, where they become reduced in size. The female is oviparous.

The verge is dorsoventrally flattened, and is attached to the nape slightly to the right of the median line. When the verge is erect it extends along the right side of the body, and passes below or to the side of the right tentacle. When it is contracted it is recurved over the middle of the nape, and points backward on the left side of the median line. A simple vas deferens extends along the right margin to the tip of the verge, where it opens beneath a small, slender apical papilla. The verge is unpigmented, except for slight shadings near the tip. Numerous nonglandular papillae are present along the margins and may also occur around the base. These papillae are arranged in three series. (1) The right margin has 7-50 papillae that are in 1-5 rows. The series begins at or near the base and extends nearly to the tip. (2) The left margin bears 1-4 papillae that are confined to the distal half of the verge. (3) An additional series may occur around the base of the verge; these are usually confined to the outer surface, but may also occur on the inner surface facing the body wall. Usually there are only a few papillae in this series, although some specimens may have 10 or more. The dextral (1) and the sinistral (2) series of papillae are characteristic of all species. The basal series (3) occur only in *L. tenuipes.*

The papillae are covered with a thin, intensely staining, single-celled layer of epithelium. They contain about 10-25 vacuoles or sinuses that repeatedly anastomose to form a labyrinth, and are continuous with the lymphatic passages of the verge. The sinuses are separated from each other and from the epithelial layer by a dense network of muscle fibers and connective tissue (Fig. 38, D). A large multiganglionated nerve chord extends along the vas deferens, and gives off a series of neurons that extend to or from the bases of the papillae.

The papillae are probably sensory structures that stimulate the male during copulation. This is suggested by their nonglandular nature, their vacuoles, which apparently can be filled with lymphatic fluids and distend the papillae, and their heavy innervation.

The shell of *Littoridinops* may generally be recognized by its elongate shape, small, protruding nuclear whorls, shallow suture, incomplete peristome, straight-edged lip in lateral profile, and paucispiral operculum with about 2 whorls. Some populations of *Littoridinops monroensis* are so variable that shell characters alone will not separate them from *Cincinnatia, Heleobops,* or *Pyrgophorus. Littoridinops* can always be recognized by the multipapillate nature of the verge, which is unique in the family Hydrobiidae. The verge is also the only structure that will consistently separate the three species recognized in this paper.

Littoridinops is known to occur in coastal areas from central Georgia to the panhandle region of Florida, and the Bahama Islands. It probably has a much wider distribution. Pilsbry (1952, 52-53) identified as *L. tenuipes* some hydrobiids from the northeastern states. His anatomical figures indicate that another species was involved, which remains unnamed.

Within Florida, *Littoridinops* is the most widely distributed hydrobiid genus. It is found along both coasts as far south as Key

West, and throughout the St. Johns drainage system. It also has a wider ecological distribution than any other genus, being found in most brackish and freshwater habitats in Florida.

Texadina Abbott and Ladd (1951, 335) is not related to *Littoridinops.* Characteristics of its verge relate it to *Heleobia* of South America (personal observation).

The species of *Littoridinops* are readily identified by the arrangement of the papillae on the verge. *Littoridinops tenuipes* (Couper) has 3-10 small papillae around the base of the verge, and a series of 7-15 papillae along the right margin, which are confined to a single row. *Littoridinops monroensis* (Frauenfeld) and *L. palustris* (new species) both lack a basal series of papillae. *Littoridinops monroensis* has 17-50 papillae along the right margin, which are arranged in three to five rows. *Littoridinops palustris* has 5-15 papillae along the right margin, which are arranged in one or two rows.

The species are difficult to distinguish by the shell due to the complex variation that occurs, particularly in *L. monroensis.* The only distinctions that can be made between *L. tenuipes* and *L. palustris* are slight differences in color and shape of the spire. The shell of *L. tenuipes* usually has a brownish hue, and the sides of the spire are slightly convex, particularly along the early whorls. *Littoridinops palustris* is olivaceous, and the spire is straight-sided. These characters are not consistent for either species. There is no doubt that they are distinct because of the nature of their verges, and their geographic isolation. *Littoridinops tenuipes* is confined to the inland waterway system along the east coast of Georgia and Florida, and *L. palustris* is confined to marshy areas near the mouth of the Withlacoochee River along the gulf coast. No populations of *Littoridinops* referable to either species has been found in intermediate areas.

Both *L. tenuipes* and *L. palustris* differ from *L. monroensis* by having a larger number of whorls/mm (Fig. 7). About 75 per cent of shells more than 3.5 mm long and with 5.5 or more whorls are separable. *Littoridinops tenuipes* has 1.49-2.26 whorls/mm; *L. palustris* has 1.49-1.87 whorls/mm; *L. monroensis* has 1.09-1.67 whorls/mm. *Littoridinops tenuipes* and *L. palustris* also differ from *L. monroensis* by the shape of the operculum. The parietal margin of the first two species is concave. *Littoridinops monroensis* has a straight or slightly convex outline to the parietal margin.

Further differences between *L. tenuipes* and *L. monroensis* are in the ratios of length/diameter (Fig. 8) and length/aperture (Fig. 9). *Littoridinops tenuipes* differs from *L. monroensis* by being slightly more attenuate and by having a proportionately smaller aperture. With respect to these two characters *L. palustris* is more like *L. monroensis* than like *L. tenuipes*. If it were not for the distinctness of its verge, *L. palustris* would have been considered as

FIGURE 7

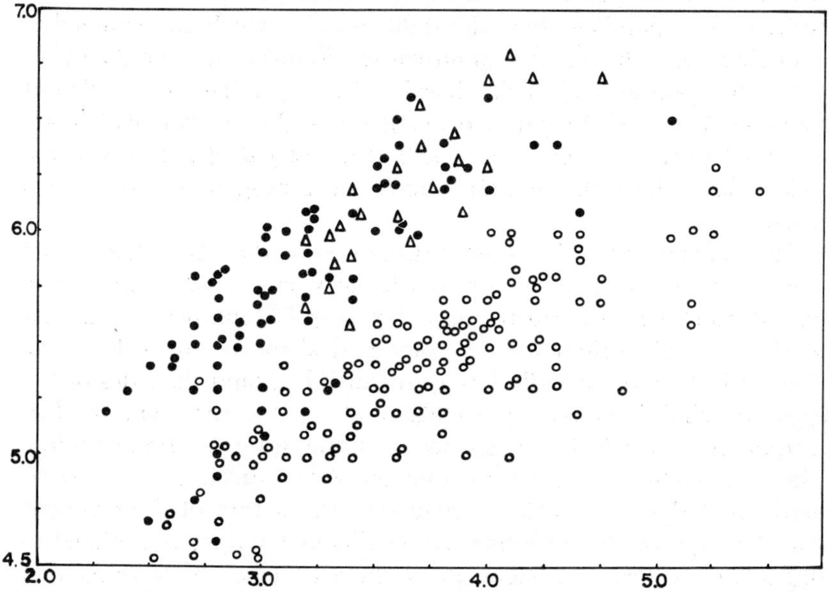

The relationship between the number of whorls and the length of the shell of three species of *Littoridinops*. Solid circles represent *L. tenuipes*, hollow circles represent *L. monroensis*, and hollow triangles represent *L. palustris*.

merely another variation of the conchologically enigmatic *L. monroensis*, which is found in the same area.

***Littoridinops tenuipes* (Couper)**

Amnicola tenuipes Couper, 1844, in Haldeman's Monograph of the Freshwater Univalve Mollusca of the United States, pt. 7, 4th page of cover. – Haldeman, 1845, *ibid.*, pt. 8: 23, pl. 1, fig. 14.

Littoridina (*Littoridinops*) *tenuipes* (Couper), Pilsbry, 1952, Naut., 66: 51-54 (in part), figs. 1, 2.

SHELL (Fig. 60, A, B; Fig. 50, A-C). Shell conical, becoming relatively narrower with age; generally thin and transparent, but may be thickened and translucent or opaque in large specimens, particularly in the last whorl or so. Light brown or olivaceous, with an oily luster. Narrowly umbilicate or rimate, occasional specimens imperforate. Spire straight-sided or slightly convex, 1-2 times the length of the aperture. Sculpture consisting of fine or vague incremental striations; spiral striations seldom evident even in thin juvenile shells. 5.0-6.6 whorls; embryonic whorls protruding, white; nuclear whorl flat-topped, 0.10-0.15 mm in diameter; remaining whorls slightly arched, nearly flat-sided. Suture distinctly impressed, but shallow. Aperture ovate; interior white in larger speci-

FIGURE 8

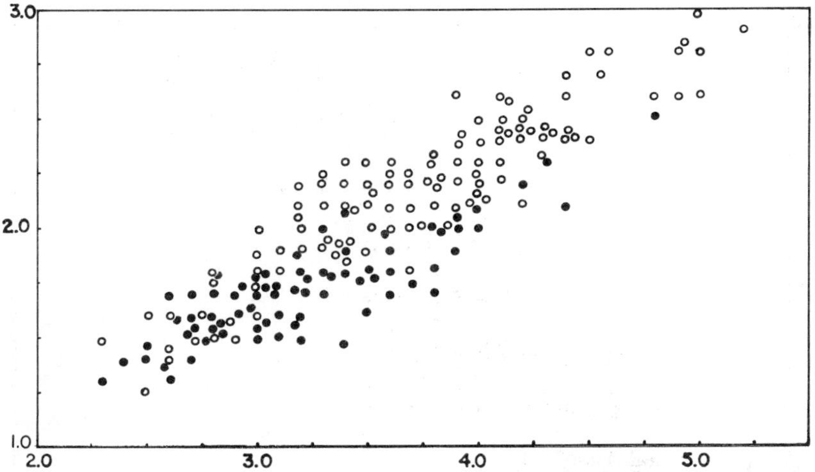

The relationship between the length of the shell and the diameter of the shell in two species of *Littoridinops*. Solid circles represent *L. tenuipes*, hollow circles represent *L. monroensis*.

mens with callused deposits; upper corner angulate, at about 90°; axis of aperture at about 20-30° to axis of shell; plane of aperture oblique, at about 20° to shell axis. Peristome continuous or not, when continuous it is adnate to the preceding whorl; outer lip sharp, continuously rounded with columellar margin, usually thickened internally, straight-sided in lateral profile; columellar lip thin, weakly arched, weakly reflected, tending to obscure rimation.

Length of shell, 2.4-4.8 mm; diameter of shell, 1.3-2.5 mm; length of aperture, 1.1-1.9 mm; width of aperture, 0.8-1.3 mm; length/diameter, 1.58-2.18; length/aperture, 2.00-3.00; whorls/length, 1.49-2.26.

Ontogenetic changes in the shape of the shell are the bases for the greatest degree of variation observed in this species. As the shell matures it becomes more attenuate, and the inside of the lip may become conspicuously thickened. These differences cause lots consisting of small shells to have a different basic appearance than lots containing predominantly large shells. No significant geographic variation has been observed.

FIGURE 9

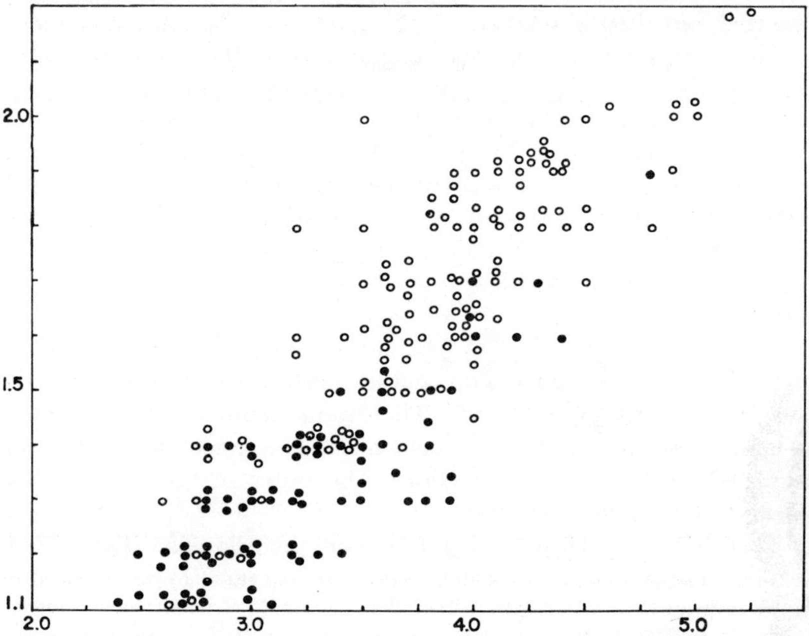

The relationship between the length of the shell and the length of the aperture in two species of *Littoridinops*. Solid circles represent *L. tenuipes*, hollow circles represent *L. monroensis*.

Operculum. (Fig. 29, C). Apex bluntly pointed; columellar margin nearly straight; parietal margin weakly concave; outer margin and basal margin evenly rounded.

Animal (Fig. 25, F). The foot, pedicel, and snout are heavily pigmented with black flecks, and a reticulated or lined pattern may be formed on the sides of the snout. The pigmentation on the snout and the nape is continuous across the dorsal surface. A narrow, unpigmented band extends from the sides of the body, within the mantle cavity, obliquely anterior to and around the anterior margin of the foot below the pedicel. In some populations the band may not be very distinct, and the color of the animals may be much lighter.

VERGE (Fig. 40, A, B). The verge is stout and about 2 mm long at maturity. The right margin bears a series of 7-15 papillae that are arranged in a single row. There are 2-4 papillae along the left margin near the center or toward the tip, and usually 3-10 papillae at and around the base of the verge. The papillae along the right and the left margins are about equal in size. Those on the base of the verge are usually smaller.

The distribution and number of papillae on the verge may vary considerably. Usually the basal series is confined to the region immediately around the base of the verge. In some specimens the papillae are more generally scattered on the basal half of the verge, and occasionally they may be widely scattered over the inner surface. The dextral series of papillae may be clumped near the middle of the right margin, or they may be more generally distributed. The sinistral series is usually located near the tip of the verge.

RADULA (Fig. 34, C). Ten specimens examined. The teeth occur in 43-52 transverse rows. The central tooth is about 35μ wide. Its cusp formula is $\frac{5\text{-}1\text{-}5}{2\text{-}2}$ to $\frac{5\text{-}1\text{-}5}{1\text{-}1}$. The lateral tooth has 2 entocones, a large mesocone, and 3 or 4 ectocones. The inner marginal tooth has 14-19 cusps on the reflection. The outer marginal tooth has 16-21 cusps on the reflection.

TYPE LOCALITY. Hopeton Plantation, on the Altamaha River about 5 miles above Darien, McIntosh County, Georgia. Type specimens not examined.

DISTRIBUTION (Fig. 10). Streams draining into the Atlantic Ocean and the Inland Waterway of East Florida and Georgia, from Dade County, Florida, north at least to McIntosh County, Georgia.

LOCALITIES. Florida: *Brevard Co.*—creek 0.3 mi. s. Grant. *Clay Co.*—Peters Creek, 3.5 mi. n.n.w. Green Cove Springs. *Dade Co.*—mangrove swamp, North Miami Beach. *Duval Co.*—creek on Fla. 13, 5 mi. s. Jacksonville; Jacksonville (*UMMZ, MCZ*). *Flagler Co.*—Pellicer Creek where U.S. 1 crosses Flagler-St. Johns county line. *Martin Co.*—Kitchen Creek, Jonathan Dickinson State Park; south fork St. Lucie R., 4 mi. s.w. Salerno. *Nassau Co.*—Lofton Creek, 1.7 mi. e. Yulee; ditch entering St. Marys R. at U.S. 17.

FIGURE 10

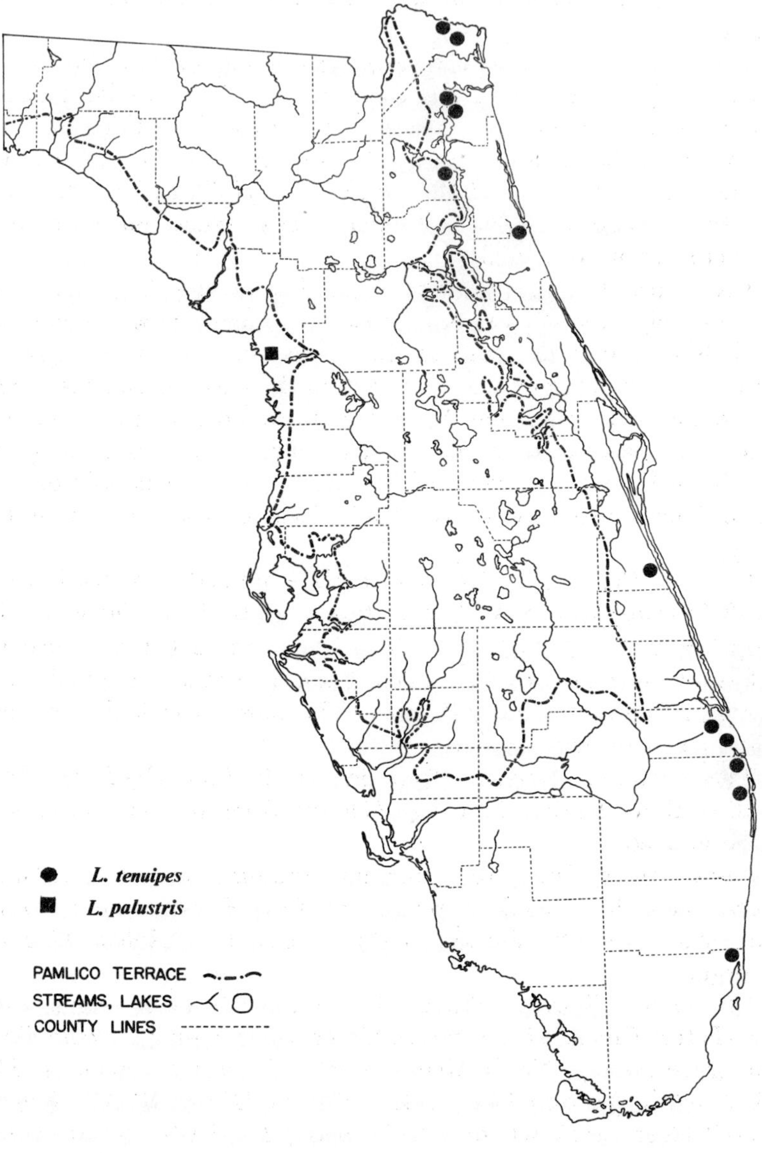

The geographic distribution of *Littoridinops tenuipes* (Couper) and *Littoridinops palustris* new species in Florida.

Palm Beach Co.—pond 1 mi. s. Jupiter; canal 3.5 mi. n.w. West Palm Beach.

Georgia: *McIntosh Co.*—canal 3.3 mi. n.n.w. Darien; ditch 3.6 mi. n.n.w. Darien; canal 4.7 mi. n.n.w. Darien; creek 2.0 mi. n. Ridgeville; creek 5.6 mi. n. Ridgeville.

ECOLOGY. Live animals were found in streams and ditches with little or no current, and usually on a hard sand bottom. Live aquatic plants were usually sparse or absent, although grasses (*Panicum* sp.) frequently formed heavy floating mats along the margins of the streams, and at one locality the creek was covered with water hyacinths (*Eichhornia crassipes*). Snails were usually found crawling on grass mats, hyacinth roots, and dead plant material that was present on the bottom of the streams or ponds.

The water in which the snails were found was usually clear, though colored by humic acids. The snail was generally found in fresh water, but was occasionally found in brackish water with a Cl^- concentration of up to 16 ppt.

Littoridinops palustris new species

SHELL (Fig. 60, C, D; Fig. 50, D-F). Elongate-conical, narrow; thin, opaque or translucent in adult shells; light olivaceous gray, with an oily luster. Narrowly umbilicate or rimate. Spire conical, straight-sided, or slightly convex, 1.1-1.8 times the length of the aperture. Sculpture consisting of distinct, relatively coarse incremental striations; very delicate, sparse, spiral striations may be evident in immature shells. 5.6-6.8 whorls; nuclear whorl weakly protruding, rounded, 0.12-0.15 mm in diameter; remaining whorls regularly descending to the aperture, slightly arched, nearly flat-sided; last whorl more strongly arched, with its periphery extending slightly beyond the slope of the spire. Suture weakly, but distinctly impressed. Aperture ovate to elliptical; white; upper corner angulate, at about 90°; axis of aperture at about 20° to axis of shell; plane of aperture oblique, at about 20° to shell axis. Peristome discontinuous across parietal margin, except in very old specimens; outer lip thin, sharp, frequently with a thick callused deposit within, occasionally dilated above periphery, continuously rounded with columellar lip, straight-edged in lateral profile; columellar margin arched, thin, weakly reflected.

Length of shell, 3.2-4.5 mm; diameter of shell, 1.5-2.1 mm;

length of aperture, 1.2-1.7 mm; width of aperture, 0.9-1.2 mm; length/diameter, 1.88-2.29; length/aperture, 2.12-2.82; whorls/length, 1.49-1.87.

Measurements of type: length of shell, 4.12 mm; diameter of shell, 2.03 mm; length of aperture, 1.56 mm; width of aperture, 1.21 mm; 6.7 whorls.

OPERCULUM (Fig. 29, D). Broadly oval, with a bluntly pointed apex, parietal margin nearly straight, slightly concave; columellar margin weakly convex; outer margin and basal margin evenly rounded.

ANIMAL. The color of the animal is similar to that of *L. monroensis,* though less intense than usual in that species. The sides of the snout are lightly pigmented, forming a large patch on each side. The patches are discontinuous middorsally. The sides of the body and the pedicel are slightly clouded by melanophores.

VERGE (Fig. 40, E-G). The verge is relatively small and slender. It is nearly uniform in width with a bluntly rounded tip. The right margin bears 9-15 papillae, which are arranged in one or two rows. The left margin of the verge has 2-3 papillae.

The papillae on the verge of this species are more variable in shape than is usual in other members of this genus. They may be conical, lanceolate, or cylindrical.

RADULA (Fig. 34, A). Five specimens examined. The teeth occur in 41-48 transverse rows. The lateral tooth is about 25-30μ wide; its cusp formula is $\frac{5\text{-}1\text{-}5}{1\text{-}1}$ to $\frac{6\text{-}1\text{-}6}{2\text{-}2}$. The lateral tooth has 2 or 3 entocones, a large mesocone, and 4 ectocones. The inner marginal tooth has 16-18 cusps. The outer marginal tooth has 20-23 cusps.

TYPE LOCALITY. Brackish marsh 3.2 miles s.w. of Yankeetown, Levy County, Florida; collected April 13, 1963 by Fred G. Thompson. Type: *UMMZ* 216492. Paratypes: *UMMZ* 216493 (150), *USNM* 636946 (75).

DISTRIBUTION (Fig. 10). The species is known only from the type locality, where it was collected in a shallow muck-bottomed pool. It probably occurs in adjacent regions along the west coast of the peninsula in ecologically similar situations, but such habitats are generally inaccessible and difficult to locate.

ECOLOGY. The pool from which the type lot was collected is located about one-half mile inland from the Gulf of Mexico, and is separated from the sea by a wide zone of rushes. The pool was 3-5 inches deep, with a muck bottom. Dense mats of aquatic angiosperms and algae filled the pool. Snails were collected by washing vegetation mats in a pail of water.

Littoridinops monroensis (Frauenfeld)

Hydrobia monroensis Frauenfeld, 1863, Verhandel Kais. Konig. Zool. Botan. Ges. Wein 13: 1023.—Pilsbry, 1899, Proc. Acad. Nat. Sci. Phila.: 87-88; pl. 3, figs. 17-19.

Bythinella monroensis (Frauenfeld) Tryon, 1870, Cont. Haldeman's Monog. Freshwater Univ. Moll. U.S.: 48 (not *Bythinella monroensis* Dall, 1885).

Littoridina monroensis (Frauenfeld) Pilsbry, 1899, Naut., 13: 22.—Walker, 1918, Misc. Publ. Museum Zool. Univ. Mich., (6): 141.

Amnicola forsythi Pilsbry, 1930, Proc. Acad. Nat. Sci. Phila., 82: 300; pl. 22, fig. 7. Type Locality: Lake Forsyth, Andros Island.

Amnicola oscitans Pilsbry, 1930. Proc. Acad. Nat. Sci. Phila., 82: 300-301; pl. 22, fig. 8. Type Locality: Lake Forsyth, Andros Island.

Amnicola sablensis Pilsbry, 1951, Naut., 64: 119-20; pl. 9, figs. 1-1c. Type Locality: West Lake. Cape Sable, Florida.

The following description of the shell applies to the typical and most widely distributed form.

SHELL (Fig. 61, A, B; Fig. 51, A). Elongate-conical; thin, transparent or translucent; periostracum light brown to light gray, with an oily luster; embryonic whorls white. Umbilicus rimate or narrowly perforate, obscured partially by reflected columellar lip. Spire conical, nearly straight-sided, only weakly convex, 1.15-1.75 times the length of the aperture. Sculpture consisting of numerous close, fine but distinct incremental striations, and finer, sparse, spiral striations, which can best be distinguished with transmitted light. 4.5-6.2 whorls; nuclear whorl strongly protruding, 0.17-0.23 mm in diameter; later whorls regularly increasing in size, weakly arched. Suture weakly or moderately impressed. Aperture elliptical to ovate; interior hyaline, may be white in large specimens; upper corner angulate, outer lip forming an angle of about 90°; axis of aperture lying at about 20° to axis of shell; plane of aperture lying at about 10° to shell axis. Peristome discontinuous across parietal margin, except in some large old specimens; outer lip thin, sharp, straight-edged in lateral profile, but extending slightly forward at insertion; outer lip evenly rounded to columellar margin, at which point it usually forms a blunt angle with columellar margin and may extend forward as a slight, rounded, tongue-like projection; columellar margin arched, nearly straight, thin, weakly reflected; parietal callus very thin, glassy, smooth, with a white tinge.

Length of shell, 3.5-4.5 mm; diameter of shell, 1.8-2.4 mm; length of aperture, 1.4-1.9 mm; width of aperture, 1.1-1.4 mm; length/diameter, 1.60-2.00; length/aperture, 2.15-2.75; whorls/length, 1.33-1.67.

Measurements of lectotype; length of shell, 3.85 mm; diameter of shell, 2.20 mm; length of aperture, 1.65 mm; width of aperture, 1.25 mm; 5.5 whorls; diameter of nuclear whorl, 0.22 mm.

This is a highly variable species that occupies a wide variety of habitats. Some of the variations are correlated with ecological changes, while others appear to be random. The most distinct variations that have been encountered are as follows.

The form described as *Amnicola sablensis* is the common brackish water variety (Fig. 51, B; Fig. 61, C, D). It differs from the typical form by having a thicker shell, and by having the peristome continuous across the parietal wall. Topotypic specimens of "*A. sablensis*" from West Lake, Everglades National Park are anatomically identical with typical specimens of *L. monroensis* from the St. Johns River, and complete intergradation of shell characters occurs between the two forms in populations transitional from brackish to fresh water. This brackish water variety is generally distributed along the east coast of Florida, through the Florida Keys to Key West, and along the west coast at least as far north and west as Wakulla County, Florida. Although it is the usual form found in brackish water, populations of typical *monroensis*, as well as other variations, may also be found in this habitat.

Specimens from ponds in the vicinity of Androstown, Andros Island, Bahama Islands completely encompass and bridge the characters used by Pilsbry to distinguish *Amnicola forsythi* and *A. oscitans*. These populations are anatomically identical with *L. monroensis*.

Population from Ft. George, Duval County. This population is like "*sablensis*," except that the suture is more deeply impressed, the shell has a thicker callus within the outer lip, and the aperture is relatively larger and more auriculate due to the shouldering of the whorls.

Population from an excavated lawn pond on Big Pine Key, three-quarters of a mile s.w. of the causeway to West Summerland Key, Monroe County (Fig. 51, F). A dwarf form with 5 whorls in which the shell seldom exceeds 2.5 mm in length. The peristome is complete, and in larger specimens the aperture may be free from the preceding whorl. The aperture is proportionally small, and the suture descends rapidly on the last whorl. The embryonic whorls are usually flattened, and the early whorls are suppressed, producing an ovate outline in some shells. Other populations on Big Pine Key represent all intermediate stages between this form, "*sablensis*," and typical *L. monroensis*.

Populations from Greynolds Park, North Miami Beach, and a pond at Matheson Hammock, Coral Gables, Dade County. A dwarf form with 6 whorls, in which the shell does not exceed 3.0 mm. The peristome is usually incomplete, the suture descends rapidly to the last whorl, and the aperture may be constricted.

Population from a quarry, 3 miles west of North Miami Beach, Dade County (Fig. 51, E). This form is like "*sablensis*," except that the aperture is completely free from the preceding whorl, and is reduced in relative size.

Population from a mangrove swamp on Middle Torch Key, sec. 18, Monroe County (Fig. 51, D). A large form commonly reaching 5.0 mm in length, and acquiring 6.3 whorls with a deeply impressed suture. Similar forms were found near Estero and Bonita Springs, Lee County, and Sulfur Springs, Hillsborough County.

Population from the Withlacoochee River below the dam at the west end of the reservoir, Citrus County (Fig. 51, C). This population is represented by a form that has a complete peristome, and about 6 closely coiled whorls with a deeply impressed suture. Intermediate and typical specimens of *L. monroensis* occur in this same population. A similar form is found near the mouth of the Peace River, DeSoto County. It differs by having heavy, irregularly spaced vertical fimbriations.

Most of these forms could be considered specifically distinct from *L. monroensis* were it not that complete intergradation occurs, and that anatomical uniformity exists throughout all of this variation. It is interesting to note that the most distinct variations encountered were found in recently dredged or disturbed habitats, such as ponds, quarries, and canals. To give nomenclatorial recognition to these populations would be license for the naming of dozens of other populations in south Florida that also have had an evolutionary age of less than twenty years.

OPERCULUM (Fig. 29, E). Apex bluntly pointed; outer margin and basal margin evenly rounded; columellar margin nearly straight; parietal margin convex, occasionally nearly straight.

ANIMAL (Fig. 25, E, G). The sides of the snout are heavily shaded with dark flecks, which may form a reticulated network of bands. The coloration is discontinuous middorsally, so that a light zone extends from the snout to the nape, beneath the mantle. The sides of the foot are clouded with melanophores. The clouding becomes darker around the anterior margin of the pedicel.

VERGE (Fig. 40, C, D). The verge is a stout structure that gradually tapers distally. The left margin bears a hump near the end, upon which are located 1-3 papillae of the sinistral series, although some of the papillae of this series may not be on the hump. The right margin becomes strongly curved near the tip, and in profile the outline of a dog's head is suggested. The dextral series is composed of 17-50 papillae, which are arranged in 3-5 rows along the right margin. The outermost row usually has the largest number of papillae, and the innermost row usually has the smallest number. The rows are usually straight, and well-defined, but on occasion the papillae may be randomly dispersed.

The number of papillae and the number of rows in the dextral series are highly variable in all series of specimens examined. The degree of variability is not correlated with geographical or ecological distribution.

RADULA (Fig. 34, B). Twenty-two specimens examined. The teeth are arranged in 43-56 transverse rows. The central tooth is 32-42μ wide; its cusp formula is $\frac{4\text{-}1\text{-}4}{1\text{-}1}$ to $\frac{5\text{-}1\text{-}5}{3\text{-}3}$. The lateral tooth has 2-1-4 cusps. The inner marginal tooth has 17-20 cusps along the reflection. The outer marginal tooth has 15-17 cusps along the reflection.

TYPE LOCALITY. Frauenfeld (1863, 1023) cited two series of specimens that he had identified as this species. One series was from Lake Monroe (Volusia County), Florida, and the other series was from Lakeman Rok, East Florida. Pilsbry (1899, 22) stated that all of Frauenfeld's specimens had come from Lake Monroe. I have not been able to locate any place name in Florida called Lakeman Rok (or Rock). I suspect that this is a phonetic misinterpretation of Lake Monroe, as it might sound to a German. Perhaps Pilsbry had additional knowledge about this matter, which he did not state.

Frauenfeld's series of specimens from Lake Monroe has not been located. The series from Lakeman Rok is in the Vienna Naturhistorisches Museum. I have selected one of these specimens as a lectotype. The remaining cotypes include 20 specimens of *L. monroensis* and one specimen of *Cincinnatia floridana.*

DISTRIBUTION (Fig. 11). This species is known from Florida and the Bahama Islands. In Florida it is generally distributed along both coasts and the Florida Keys. It is found primarily in brackish water, but has invaded marginal freshwater, and occurs throughout the St. Johns drainage system. It has not been found far inland in any of the drainage systems along the west coast of the state, nor in any system along the east coast other than the St. Johns. *Littoridinops monroensis* may be considered a species that originally occupied brackish and marginal fresh water habitats. Its presence in the St. Johns systems is probably a relict distribution from a time during the Pleistocene when sea levels were higher. During such times the St. Johns River basin would have been a large bay from the Atlantic Ocean.

In the Bahama Islands it has been recorded as *Littoridina tenuipes* from Grand Bahama Island (Clench, 1938: 314), Crooked Island and Acklins Island (Clench, 1963: 402), and South Bimini Island (Parodiz, 1957: 2). In addition I have examined specimens from Andros Island and Cat Island.

LOCALITIES. Florida: *Brevard Co.*—creek 1 mi. s.w. Melbourne; 2 mi. e. Titusville; ditch 4 mi. w. Wilson; ditch 3.5 mi. n. Wilson; pond 7 mi. w. Mims. *Broward Co.*—canal, Ft. Lauderdale Power Plant. *Charlotte Co.*—pond 5.8 mi. s.s.w. McCall; marsh 6.4 mi. s.s.w. McCall. *Citrus Co.*—Chassahowitzka R., Chassahowitzka; Homosassa Springs; Crystal River; Withlacoochee R. below dam to reservoir. *Clay Co.*—St. Johns R., Orange Park. *Collier Co.*—canal 10 mi. e. Monroe Station; canal by x-section U.S. 41 and Fla. 29; canal 1.3 mi. w.n.w. x-section U.S. 41 and Fla. 29; canal 8 mi. w.n.w. x-section U.S. 41 and Fla. 29; mangrove swamp 0.3 mi. n. Everglades; pond 1.5 mi. n. Everglades; pond 1.8 mi. n. Everglades; canal 2.4 mi. n. Everglades; pond 2.7 mi. n. Everglades; canal 0.7 mi. e. Ochopee. *Dade Co.*—Greynold Park, North Miami Beach; quarry 3 mi. w. North Miami Beach; head Miami River (*UMMZ*); canal along Palmetto Bypass, N.W. 123 St.;

FIGURE 11

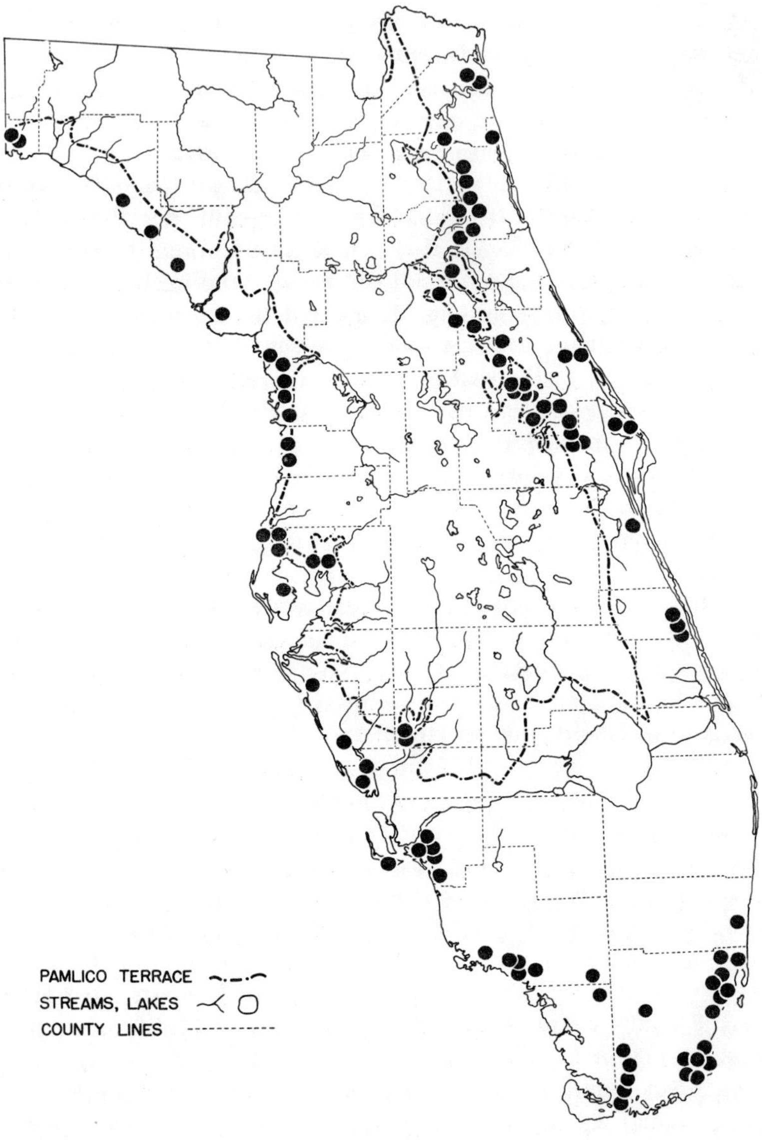

The geographic distribution of *Littoridinops monroensis* (Frauenfeld) on the Florida mainland.

Crandon Park, Key Biscayne; Matheson Hammock, Coral Gables; canal 7 mi. e. Homestead; ditch 8.2 mi. s.w. Florida City; canal 1.7 mi. n.w. Card Sound; canal 4.4 mi. n.w. Card Sound; canal 5.6 mi. n.w. Card Sound; canal 6.5 mi. n.n.w. Card Sound; Everglades National Park, canal along 7-Mile Tower Rd., 3.5 mi. s. U.S. 41. Everglades National Park, from numerous localities between Flamingo and Mahogany Hammock, about 17 mi. inland. *DeSoto Co.*—Peace R., 1.9 mi. w.n.w. Fort Ogden; Peace R., 2 mi. w. Fort Ogden; Peace R., 3 mi. s.w. Fort Ogden. *Dixie Co.*—canal 0.4 mi. n. Horseshoe Beach. *Duval Co.*—marsh 8.1 mi. w. Ft. George; marsh 0.6 mi. n. Ft. George. *Hernando Co.*—Weeki Wachee Spring Run, 1 mi. s.e. Bayport; spring run 2 mi. s.s.e. Bayport. *Hillsborough Co.*—Sulphur Springs; Hillsborough R., Tampa. *Indian River Co.*—canal 3.2 mi. s. Vero Beach; canal, n. edge Vero Beach; creek 3.2 mi. n. Vero Beach. *Lake Co.*—St. Johns River, 5 mi. w. Deland; St. Johns R., Astor; Lake Yale, 3 mi. n. Grand Island. *Lee Co.*—canal, Bonita Springs; creek 1 mi. n. Estero; creek 3.7 mi. e. intersection Fla. 865-Fla. 867; creek 1.1 mi. e. intersection Fla. 865-Fla. 967; creek 1.1 mi. s.w. Ft. Myers; ditch 10.4 mi. s.w. Ft. Myers; Turners Slough, Sanibel Isl. (*MCZ*). *Levy Co.*—marsh 1.8 mi. s.w. Yankeetown; pond 0.6 mi. s.s.w. intersection Fla. 24-Fla. 347. *Marion Co.*—Salt Springs; Silver Glen Springs. *Monroe Co.*—several localities on Big Pine Key; Middle Torch Key; Ramrod Key; Summerland Key; canal, Loop Road, 7 mi. w. Tamiami Trail at 40-Mile Bend. *Orange Co.*—creek 2.1 mi. e. Christmas; St. Johns R., 4 mi. e. Christmas. *Pinellas Co.*—St. Petersburg (*UMMZ*); north and south ends of Lake Tarpon; marsh 1.3 mi. n. Tarpon Springs. *Putnam Co.*—creek, East Palatka; St. Johns R., East Palatka; Welaka Springs, 1 mi. n. Welaka; slough 2 mi. n. East Palatka; Fish Creek, 3.3 mi. s. Bostwick. *St. Johns Co.*—river 0.2 mi. s. Spuds; creek 3.1 mi. n.n.w. Spuds; creek 4.0 mi. s. Tocoi; creek 0.7 mi. s. Tocoi; creek 6.5 mi. n. Tocoi; pond 2.3 mi. w. Ponte Vedra Beach. *Sarasota Co.*—canal 1.6 mi. s.w. Fruitville; canal 2 mi. n.w. Port Charlotte; Warm Mineral Springs (*MCZ*). *Seminole Co.*—creek 7.4 mi. s.e. Geneva; St. Johns R., 5.4 mi. s.e. Geneva; Lake Harney near Osceola; Lake Jessup, 2 mi. n. Oviedo; north end Lake Jessup at Fla. 46. *Taylor Co.*—ditch, Steinhatchee; marsh 0.6 mi. s.e. Keaton Beach. *Volusia Co.*—St. Johns R., 4 mi. e. Sanford; Green Springs, 1 mi. s.e. Enterprise; Lake Monroe; Padgett's Spring, 1 mi. e. DeBary; creek 3.3 mi. w. Brevard County line on Fla. 46; Lake Woodruff, De Leon Springs; New Smyrna Beach; 2 mi. w. New Smyrna Beach. *Wakulla Co.*—St. Marks R., Newport; Wakulla R. at U.S. 98.

Bahama Islands: *Andros Island*—Androstown; pond 0.5 mi. s.e. Androstown; pond 1.5 mi. s.e. Androstown. *Cat Island*—pond 1 mi. s. Tea Bay; pond 0.5 mi. n. Arthurs Town.

ECOLOGY. *Littoridinops monroensis* is euryhaline, being found abundantly in brackish and fresh water. The highest salinity tested from field samples had a Cl^- concentration of 19.98 ppt. The species undoubtedly can tolerate a greater salinity than this, for it has been found in tidal marshes that nearly dry up due to evaporation. Population size appears to be independent of salinity.

Specimens were most commonly found on aquatic vegetation, such as *Elodea, Ranunculus, Panicum, Utricularia, Eichhornia, Chara,* and mats of filamentous algae. They were also common in diatomaceous muck and ooze. On rare occasions they were found on sticks, stones, or bare sand. Usually this species was found in clear water, but it can withstand very turbid conditions.

Aphaostracon new genus

The sexes are dimorphic. The female is larger, and there are usually more females than males in a colony. Females, because they are the most frequently collected and the most consistent in shell characteristics, are used as the basis for defining the species.

Female shell: small, 2.5-3.0 mm long or less; conical, ovate-conical, or terrete, 1.26-2.09 times as long as wide; varying from thin, fragile, and transparent to thick, solid, and opaque. Spire conical or cylindric-conical, 1.0-1.7 times the height of the aperture. Umbilicus variable in size, one-sixth or less the diameter of the shell, frequently reduced to a narrow chink. Color variable, light gray to dark rusty brown; nuclear whorls usually lighter colored than rest of shell; occasionally the periostracum may be covered with a thin dark mineral deposit. Sculpture very weak or absent; when present it consists of very fine incremental striations. 4.0-5.3 whorls; nuclear whorl very small, 0.20-0.25 mm in diameter, protruding; remaining whorls variable in contour. Suture variable in depth. Aperture elliptical, ovate, or nearly circular; interior usually hyaline, but may be livid white in some species; plane of aperture parallel with or at a slight angle to the axis of the shell. Peristome complete across parietal region at maturity; adnate to or free from preceding whorl. Outer lip thin and sharp; columellar margin thin, usually directed forward, but may be slightly reflected.

Males are usually much smaller than are females, and have about one whorl less. The aperture is proportionally larger and broader, and is almost always free from the preceding whorl by the time the shell has acquired three whorls. The shells of both sexes are similar in other details.

The operculum is thin, chitinous, colorless, and ovate or circular in shape. It consists of 3-6 whorls and varies from paucispiral to nearly multispiral. The outer surface may be sculptured with a few weak incremental striations or it may be more rugose.

The radula is taenioglossate with 40-59 transverse rows of minute teeth. The central tooth has a large mesocone, and 4-5 ectocones and a single basocone on each side of the face. The lateral teeth are clavate in shape with 2-3 entocones, a large mesocone, and 5-6 ectocones. Each of the marginal teeth has about 20 minute cusps along the rasping margin.

Females are oviparous. The verge of the male is a structurally simple organ which bears 0-6 short conical papillae along the right margin, and may also have a large lateral bulge at the base of the left margin. The verge originates on the nape slightly to the right of the middorsal line at a distance behind the tentacles about equal to the length of the snout. When contracted it is recurved to the left side beneath the mantle with its distal end pointing backward. When erect it passes along the right dorsolateral margin of the body beside the tentacle. The verge is slightly compressed dorsoventrally, but only to the extent that it is broadly elliptical in cross section. The penis is long and slender and gradually tapers to a fine point. The vas deferens is simple and lies along the right margin of the verge.

The body (*A. rhadinus*) has no striking peculiarities in color or shape. The foot is small and spatulate with a truncate anterior margin. The tentacles are about one and one-half times as long as the snout and bear a diffuse dark band near the tips. The band is continuous on both sides of the tentacle with a narrow dark stripe that extends to the base. The tip of the snout is white. The rest of the snout, nape, pedicel, and mantle are diffusely mottled or reticulated with gray. The gill lamellae lie in a long narrow series that is nearly uniform in width and extends from the middle of the mantle collar posteriorly and to the left to the anterior margin of the pericardium. The series contains about 25-28 lamellae.

Type species: *Aphaostracon rhadinus* new species.

The generic name is derived from the Greek ἀφάεος, meaning dull, and ὄστρακον, meaning little shell. It is neuter.

The shell is distinguished by its minute size, its conical or turreted shape, its minute protruding nuclear whorls, and its complete peristome in which the columellar margin is thin and directed forward or only very slightly reflected. The shell is not particularly similar to any other genus of Florida hydrobiid, and is distinguished from all by its minute size and its delicate appearance. The relationship of *Aphaostracon* to other snails of the eastern states that have been placed in *Fontigens* is not known because the verges of these other species have not yet been described.

Aphaostracon is placed in the *Pyrgophorus* tribe because of the nature of the papillae on the verge. It differs from all other genera of this tribe by having the papillae, when present, confined to the right margin of the verge. It is similar to *Littoridinops,* and differs from *Pyrgophorus* and *Hyalopyrgus* in that the females are oviparous.

Aphaostracon is generally distributed and common throughout the Florida peninsula and the panhandle at least as far west as the Wakulla River. It probably occurs in other areas, but has been overlooked because of the minute size of its species. The species are confined to fresh water where they are ecologically generally distributed.

Aphaostracon includes nine described species, which are remarkably alike in size and general appearance. Collectively, the shells are confusing to identify because of similarities of conchological features in species that are anatomically quite different. The species are divided into three groups based upon the number of papillae on the verge. All of the species within any group are readily distinguished from each other by shell characteristics. The species groups probably are not entirely natural assemblages, for it appears that the number of papillae on the verge can be easily changed through new acquisitions or loss. However, some of these groups have geographic distributions that suggest common relationships. Species with one or two papillae are probably the most generalized, and probably gave rise to forms with more or fewer papillae.

The three species groups are separated as: (1) those that have one or two papillae on the verge; (2) those that lack papillae on the verge; and (3) those with three or more papillae. The first two groups are probably natural assemblages. The last group probably has been independently derived from group (1) through two or more lineages.

Several additional species of *Aphaostracon* occur in Florida. Anatomical material sufficient for determining the relationships of these other species is not available, and so their placement within the genus has not been established. They are left undescribed because of their undetermined relationships.

The *rhadinus* Group

This group of three species is confined to the lower St. Johns drainage system and the Suwannee River system. It may also include the undescribed species from the Withlacoochee River system, the lakes of Lake and Sumter Counties, and the panhandle region. The group is characterized by one or two papillae on the right margin of the verge. The number of papillae is not constant, for there are individuals in some populations of *A. rhadinus* and *A. hypohyalina* that possess one or two papillae, although specimens with only one are the most frequently encountered. The three species may be distinguished as follows.

A. rhadinus—The shell is narrow, terrete, or cylindric-conical with a moderately impressed suture. It has 4.6-5.3 whorls which are weakly or moderately arched. The ovate aperture is adnate to the preceding whorl across the parietal margin. The operculum is paucispiral with about 3 whorls.

A. hypohyalina—The shell is conical or ovate-conical, with a deeply impressed suture. It has 4.0-4.5 broadly arched, moderately shouldered whorls. The broadly elliptical aperture may or may not be free from the preceding whorl. The operculum is paucispiral with about 3 whorls.

A. monas—The shell is elongate-conical with a deeply impressed suture. It has 4.0-4.3 whorls that are strongly arched but weakly shouldered. The broadly ovate aperture is usually free from the preceding whorl. The operculum is nearly multispiral with about 6 slowly expanding whorls. It is distinguished from all other species of the genus by this last character.

Aphaostracon rhadinus new species

SHELL (Fig. 62, A, B; Fig. 52, A-C, K). FEMALE: Shell terrete or cylindric-conical, 1.62-2.09 times as long as wide; moderately thin, but not fragile; translucent, occasionally opaque. Spire cylindric-conical or conical, sides moderately arched, 1.25-1.79 times the height of the aperture. Umbilicus rimate or very narrowly perforate; perforation nearly obliterated by the overlying peristome. Periostracum brown or dark olive-brown; nuclear whorls grayish. Sculpture consisting of fine or moderate, generally discontinuous incremental striations. 4.6-5.3 whorls. Nuclear whorl protruding, 0.20-0.24 mm in diameter. Remaining whorls moderately arched but not shouldered, regularly increasing in size. Suture moderately impressed. Aperture elliptical-ovate in shape; upper corner usually rounded at maturity, but may be slightly angulate; aperture 0.67-0.79 times as wide as long; interior hyaline or tinged with livid white. Plane of aperture at about a 10-15° angle to the axis of the shell. Peristome complete across parietal margin at maturity; adnate to the preceding whorl. Outer lip sharp-edged, moderately thin but not fragile; very slightly sinuous in lateral profile. Columellar margin sharp-edged, slightly thicker than outer lip and directed forward. Columellar margin and parietal margin gently curved from the basal lip to the upper corner of the aperture.

Length, 2.40-3.35 mm; diameter, 1.40-1.75 mm; height of aperture, 1.00-1.25 mm; length of aperture, 1.05-1.30 mm; width of aperture, 0.75-0.90 mm; length/aperture, 2.25-2.76; 1.60-2.00 whorls/mm.

Measurements of the type: length, 3.30 mm; diameter, 1.70 mm; height of aperture, 1.25 mm; length of aperture, 1.30 mm; width of aperture, 0.90 mm; 5.3 whorls; nuclear whorl 0.23 mm in diameter.

This species shows little intraspecific variation other than size, and most series of specimens are very much alike in the shape of the shells and relative proportions. Occasional samples are notable in that the specimens are relatively small and have fewer whorls than as given above. In all other respects these specimens are normal.

MALE: Males differ from females in that they generally have a conical shell with about one whorl less, they are about 1 mm shorter, and they are relatively broader. Males of this species are unusual in that the peristome is almost always in contact with the preceding whorl, whereas in males of most other species of the genus the peristome is generally separated. Measurements are: length, 1.35-1.60 mm; diameter, 0.95-1.05 mm; height of aperture, 0.65-0.75 mm; length of aperture, 0.70-0.80 mm; width of aperture, 0.50-0.60 mm; length/diameter, 1.28-1.52; length/aperture, 1.93-2.15; 2.50-2.71 whorls/length; aperture 0.72-0.79 times as wide as long.

OPERCULUM (Fig. 30, A). Ovate, apex bluntly rounded, parietal margin nearly straight; thin, chitinous, colorless, transparent. Paucispiral, consisting of about 3 rapidly expanding whorls; nucleus located at about one-third of the distance from the left margin to the right, and from the base to the apex. Outer surface sculptured with rather coarse incremental striations and fimbriations. Smooth inner surface.

VERGE (Fig. 41, A, B). The verge is a simple, elongate structure that bears a moderate bulge along the left margin near the base, and one or two small conical papillae along the right margin slightly below the middle. The penis is simple and tapers to a slender point.

The males of most populations have only a single papilla along the right margin. In most other populations in which either 1 or 2 papillae occur, specimens with a single papilla predominate. Males of the population from Welaka Springs, which are tentatively referred to this species, differ from males of other populations by having 2 or 3 papillae along the right margin.

RADULA (Fig. 33, C). The radula contains 48-59 transverse rows of teeth. The central tooth is very small and modified trapezoidal in shape, with a basolateral winglike projection on each side, and a large bulge along the basal margin. It bears 4-5 ectocones, a larger mesocone, and a single basocone on each side of the face. The lateral tooth is clavate in shape with 2-3 entocones, a large mesocone, and 5-6 ectocones. The inner marginal tooth has a broad shaft, and about 20 small cusps along the rasping margin. The outer marginal tooth has a more slender tapered shaft and about 20 very minute cusps along the rasping margin.

TYPE LOCALITY. Fish Creek where it is crossed by Florida State Highway 209, 3.3 miles east of U.S. Highway 17, 3.7 miles south of Bostwick, Putnam County, Florida. Type: *UMMZ* 216525; collected March 21, 1963 by Fred G. Thompson. Paratypes: *UMMZ* 216526

(200), *USNM* 636944 (100), *MCZ* 258008 (50), *ANSP* 301546 (30), *CM* 65-16 (57), *UF* (50).

DISTRIBUTION (Fig. 12). Known only from a small area in northeastern Florida, where it has been found in small streams and sloughs draining into the St. Johns River north of Palatka, and in a small artificial lake near the coast in St. Johns County.

LOCALITIES. Florida: *Putnam Co.*—creek at east Palatka; Welaka Springs, 1 mi. n. Welaka. *St. Johns Co.*—river 0.2 mi. s. Spuds; creek 4.0 mi. s. Tocoi; creek 0.7 mi. s. Tocoi; creek 6.5 mi. n. Tocoi; Trout Creek at Fla. 13 and Duval Co. line; artificial lake on Fla. Hwy. 210-A, Ponte Vedra Beach.

ECOLOGY. This species inhabits quiet and usually shallow streams and sloughs that drain low hardwood and cypress swamps. At most stations specimens were very abundant on floating vegetation, particularly hyacinths. They were seldom found on the bottom or on submerged objects. Water turbidity and temperature were highly variable.

Aphaostracon hypohyalina new species

SHELL (Fig. 62, C, D; Fig. 52, D-F, I). FEMALE: Small, conical, or ovate-conical due to basal position of last whorl, 1.38-1.66 times as long as wide; thin, fragile, subtransparent or translucent. Spire conical, 1.16-1.65 times as long as height of aperture; sides of spire noticeably convex. Umbilicus usually narrowly rimate due to the superimposed peristome. Periostracum dull pale olive-brown, generally free from mineral deposits. Sculpture consisting of numerous very fine continuous incremental striations. 4.0-4.5 whorls; nuclear whorl 0.20-0.23 mm in diameter, light gray in color, protruding but nearly flat-topped; remaining whorls strongly arched, but tending to be somewhat flattened below the periphery, moderately shouldered; body whorl becoming basal in position, thus obscuring the umbilicus. Suture deeply impressed. Aperture broadly elliptical, 0.72-0.84 times as wide as long; interior hyaline. Plane of aperture at about a 10-15° angle to axis of shell. Peristome continuous across parietal margin, usually narrowly separated from preceding whorl but may be more broadly separated in older specimens or adnate in some shells. Outer lip thin, slightly sinuous in lateral profile. Columellar margin thin, directed forward.

Length, 2.00-2.40 mm; diameter, 1.25-1.55 mm; height of aper-

FIGURE 12

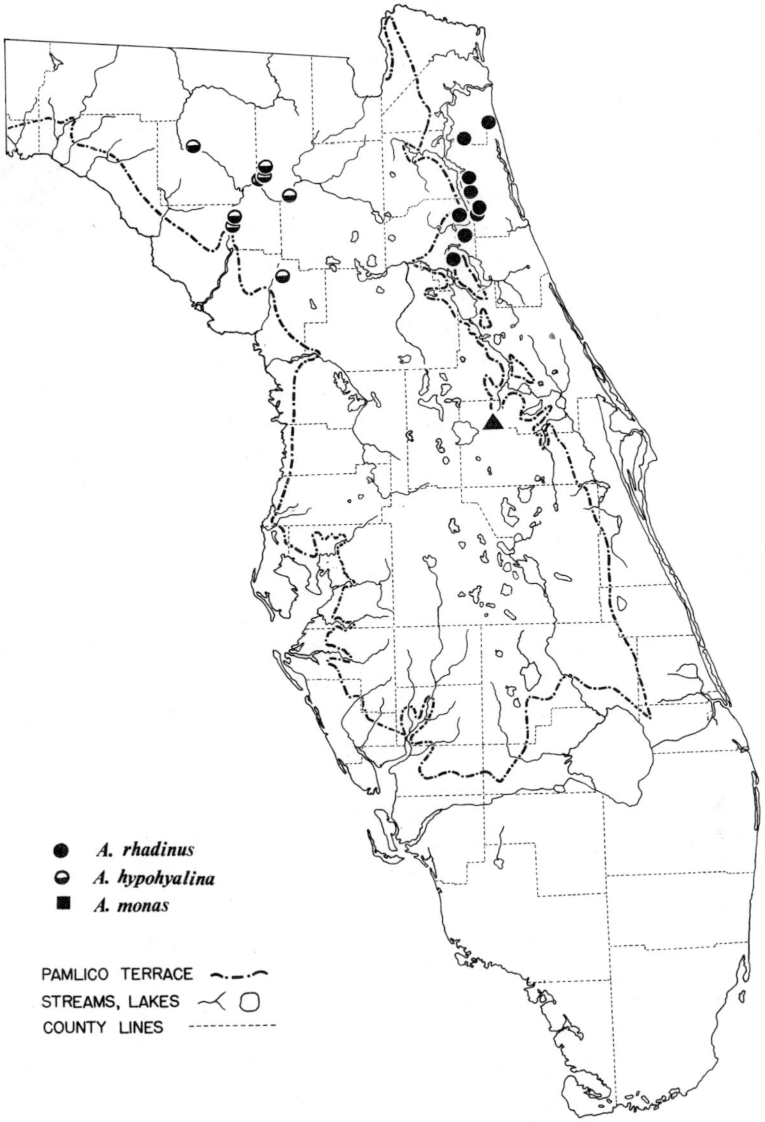

The geographic distribution of *Aphaostracon rhadinus* new species, *Aphaostracon hypohyalina* new species, and *Aphaostracon monas* (Pilsbry).

ture, 0.80-1.00 mm; length of aperture, 0.85-1.05 mm; width of aperture, 0.70-0.85 mm; length/aperture, 2.16-2.65; 1.82-2.10 whorls/mm.

Measurements of the type: length, 2.30 mm; diameter, 1.49 mm; height of aperture, 0.94 mm; length of aperture, 1.00 mm; width of aperture, 0.87 mm; 4.3 whorls; nuclear whorl 0.23 mm in diameter.

Little variation occurs within this species. Occasional old females may attain a length of up to 2.75 mm, but they are typical in all other respects. Specimens from Rock Bluff Springs, Gilchrist County, differ by having a darker brown shell and by having the outer lip nearly straight-edged in lateral profile.

MALE: Males develop a complete peristome at an earlier stage than do females, but apparently they ultimately attain the same size. Because of its early development, the peristome is more frequently separated from the preceding whorl in small males than in females of the same size. This occurs when the male reaches a length of about 1.2 mm and has about 3 whorls. In all other respects and in larger specimens, the shells of both sexes are very much alike.

OPERCULUM (Fig. 30, B). Broadly ovate, apex bluntly pointed or rounded; thin, transparent, colorless, chitinous. Paucispiral, nearly subspiral; consisting of about three and one-half whorls. The central 2 whorls very slowly increase in diameter, but the last whorl rapidly expands to a much larger size. Outer surface sculptured with a few rather distinct incremental striations. Smooth inner surface.

VERGE (Fig. 41, D, E). The verge is rather stocky when contracted and is decidedly more slender when extended. It does not possess any conspicuous enlargement near the base of the left margin. The right margin bears 1 or 2 short, conical, obese papillae, which are located much closer to the base of the verge than in most other species of this genus.

TYPE LOCALITY. Mill Spring, Ichetucknee Springs, Columbia County, Florida. Type: *UMMZ* 216527; collected March 19, 1963 by Fred G. Thompson. Paratypes: *UMMZ* 216528 (75), *UMMZ* 216529 (60), MCZ 258009 (30), *ANSP* 301541 (30), *USNM* 636941 (50), *CM* 65-15 (26), *UF* (30).

Ichetucknee Springs consists of a complex of seven large springs and many smaller ones which give rise to a large, clear, cool, fast-flowing spring run. Mill Spring is located along the southeast side of the run about one mile below the head of the system. The spring emerges from several small boils that form a quiet circular pool in the surrounding limestone.

DISTRIBUTION (Fig. 12). This species is known only from north-central Florida, where it has been found in springs along the lower half of the Suwannee River and its tributary, the Santa Fe River, and in a landlocked spring nearby.

LOCALITIES. Florida: *Alachua Co.*—Hornsby Springs, 3 mi. n. High Springs. *Columbia Co.*—Mill Spring, Ichetucknee Springs; Lingham Spring #5, Ichetucknee Springs. *Gilchrist Co.*—Rock Bluff Springs, 5 mi. n.w. Bell. *Levy Co.*—Blue Springs, 3 mi. w. Bronson. *Suwannee Co.*—small spring along the right bank of Ichetucknee Spring Run, 1 mi. n.e. U.S. 27; Telford Springs, Luraville.

ECOLOGY. This species has been found only in and adjacent to springs and spring runs. In such situations the water is nearly uniform in temperature throughout the year, and is characteristically high in mineral content. Specimens were found in a variety of microhabitats including floating vegetation, such as hyacinths and water lettuce, on tree rootlets, on submerged angiosperms, and on bottom silt and sands. Populations were usually very limited in their distributions, although the number of individuals within a colony may be very high per unit area.

Aphaostracon monas (Pilsbry)

Paludestrina monas Pilsbry, 1899; Naut., 13: 21.—Walker, 1918; Misc. Publ. Museum Zool. Univ. Mich., (6): 138.

SHELL (Fig. 62, E, F; Fig. 52, G, H, J). FEMALE: Shell small, elongate-conical, 1.53-1.64 times as long as wide; thin, fragile; translucent or barely transparent. Spire conical, slightly convex in outline, 1.26-1.63 times length of aperture. Umbilicus rimate. Periostracum usually with a thin, dull, lusterless deposit. Cleaned shells light grayish brown, with a dull shine. Sculpture consisting of very fine indistinct incremental striations that may be obscured or obliterated by rusty deposits on shell. 4.0-4.3 whorls in females; nuclear whorl protruding, rounded, 0.25-0.27 mm in diameter. Remaining whorls regularly increasing in size, very convex, weakly shouldered. Suture deeply impressed. Aperture ovate, 0.77-0.85 times as wide as long; upper corner broadly rounded; plane of aperture at about 10° to axis of shell. Peristome complete across parietal region at maturity, usually free from preceding whorl. Outer lip sinuous, lower half slightly extended forward. Outer lip and basal lip very thin and sharp-edged. Columellar margin thin, weakly reflected.

Length, 2.15-2.55 mm; diameter, 1.40-1.60 mm; height of aperture, 0.90-1.05 mm; length of aperture, 0.98-1.17 mm; width of aperture, 0.82-0.98 mm; length/aperture, 2.26-2.63; 1.68-1.83 whorls/mm.

The most conspicuous variations involve the relative obesity of the shell. Most individuals are stocky in shape. A few specimens are slender and more attenuate.

MALE: Males are considerably smaller than females. In most other respects they are similar to the female except that they are relatively more stocky, are about 1 mm shorter, and have one whorl less.

Length, 1.40-1.45 mm; diameter, 1.05 mm; height of aperture, 0.70 mm; length of aperture, 0.78 mm; width of aperture, 0.63-0.65 mm; 3.2-3.3 whorls; (2 specimens examined).

OPERCULUM (Fig. 30, C). Broadly ovate, relatively thick, chitinous, hyaline, dark amber yellow. Multispiral, consisting of about 6 tightly coiled whorls that slowly increase in size. Outer surface with numerous intense incremental striations. Inner surface smooth, flat.

VERGE (Fig. 41, C). The verge has only a single papilla on the right margin at about one-third of the distance from the base to the apex. The left margin is broadly expanded from the base to a point opposite the papilla. The penis is cylindrical and tapers to a fine point. The verge apparently is about 0.75 mm long when erect. Only a single specimen was available for examination and its verge was contracted beneath the mantle.

TYPE LOCALITY. Wekiva River, Orange County, Florida. Lectotype by present designation: *ANSP* 73907 (26 cotypes also in the *ANSP* collections); collected in February, 1886 by C. E. Beecher. The lectotype is an immature female.

DISTRIBUTION (Fig. 12). This species is confined to Wekiwa Springs and a small part of the Wekiva River for a distance of about 1 mile below the springs.

LOCALITIES. Florida: *Orange Co.*—Wekiwa Springs (four series); Wekiva River (*ANSP, MCZ*).

ECOLOGY. Snails appeared to be generally distributed on submerged plants, rocks, and gravel in and around the area of the springs; they were never abundant. Specimens were most easily collected by shaking large quantities of aquatic vegetation in a pail of water. A general description of Wekiwa Springs is given in the discussion of *Cincinnatia wekiwae*.

The *pycnus* Group

This group contains three species and is characterized by a verge that is devoid of papillae. The group is confined to springs along the middle St. Johns River system, and each species is confined to a single spring. The species are distinguished by the following characters.

A. *pycnus*—The shell is squat, compact, cylindric-conical in shape, and is usually thick and opaque. It has 4.2-4.6 weakly arched whorls with a weakly impressed suture. The body whorl is ponderous. The aperture is auriculate and is tightly appressed against the preceding whorl.

A. *theiocrenetus*—The shell is elongate-conical, thin, and subtransparent. There are 4.5-4.9 strongly arched whorls with a deeply impressed suture. The body whorl is inflated. The aperture is elliptical and is loosely attached to or slightly separated from the preceding whorl.

A. *asthenes*—The shell is small and conical. It is very thin, very fragile, and transparent. There are 4.1-4.5 strongly rounded whorls with a deeply impressed suture. The body whorl is not noticeably enlarged. The aperture is elliptical in shape or some modification thereof, and is usually attached to the preceding whorl.

Aphaostracon pycnus new species

Shell (Fig. 63, A, B; Fig. 53, A-C, J). female: Small, squat, ovate-conical, or conical, 1.56-1.80 times as long as wide; thick, translucent, or opaque at maturity. Spire compact, conical, decidedly convex in outline, 1.00-1.38 times length of aperture. Umbilical perforation less than one-twelfth the diameter of shell, partially obscured by reflected columella. Periostracum light gray or grayish brown with a dull oily luster. Sculpture consisting of very fine, indistinct, discontinuous, incremental striations. 4.2-4.6 whorls. Nuclear whorl somewhat depressed but not planular, 0.23-0.25 mm in diameter. Remaining whorls weakly or moderately arched, never shouldered; body whorl ponderous, becoming basal in position resulting in the tendency for the shell to be ovate in shape. Suture weakly or moderately impressed, and may descend slightly near aperture. Aperture ovate or auriculate, about 0.67-0.79 times as wide as long. Plane of aperture nearly parallel to axis of shell, at about a 5° angle. Peristome complete across parietal margin at maturity; tightly appressed to preceding whorl. Outer lip sharp-edged, thick, particularly at upper corner, nearly straight in lateral profile. Columellar lip partially appressed over umbilical region, usually not reflected but may be directed forward.

Length, 2.20-2.80 mm; diameter, 1.40-1.70 mm; height of aperture, 1.03-1.30 mm; length of aperture, 1.05-1.35 mm; width of aperture, 0.80-0.95 mm; length/aperture, 2.00-2.38; 1.62-1.90 whorls/mm.

Measurements of type: length, 2.70 mm; diameter, 1.60 mm; height of aperture, 1.25 mm; length of aperture, 1.30 mm; width of aperture, 0.95 mm; 4.5 whorls; diameter of nuclear whorl, 0.23 mm.

The most significant variations involve the depth of the suture and the compactness of the spire. In no instance is the suture more than moderately impressed, although it usually is not as shallow as it is in the type. The spire generally is relatively short and compact due to the shallow suture, but in more attenuate specimens it may be more loosely coiled and consist of more rounded whorls.

MALE: Only a single male was collected and apparently it is incomplete in shell development for the peristome is not continuous across the parietal margin. It is similar to female shells except for the differences of size. Its measurements are as follows: length, 1.65 mm; diameter, 1.20 mm; height of aperture, 0.90 mm; length of aperture, 0.93 mm; width of aperture, 0.63 mm; 3.8 whorls.

OPERCULUM (Fig. 30, D). Ovate, chitinous, thin, membranous, hyaline, colorless or grayish. Paucispiral, consisting of about 3 rapidly expanding whorls. Nucleus located at about two-fifths of the distance from the left margin to the right, and at about one-third of the distance from the base to the apex. Outer surface sculptured with a few fine incremental striations. Smooth inner surface.

VERGE (Fig. 41, F). The verge is elongate and slender, and lacks papillae along either margin. The basal third of the left margin has a large lateral bulge. The penis is slender and gradually tapers to a fine point. The erect verge is about 1 mm in length.

TYPE LOCALITY. Alexander Springs Run, Lake County, Florida. The species has been collected in the spring run only in the immediate vicinity of where it is crossed by Florida State Highway 445. Type: *UMMZ* 216512; collected April 22, 1962 by Fred G. Thompson. Paratypes: *UMMZ* 216513 (24), *UMMZ* 216514 (14).

DISTRIBUTION (Fig. 13). Known only from the type locality.

ECOLOGY. Specimens were collected from shallow, quiet pools along the spring run by shaking water lettuce and hyacinths in a pail of water. They were found over a soft ooze bottom in clear water that contained large patches of varied aquatic plants.

Aphaostracon theiocrenetus new species

SHELL (Fig. 63, E, F; Fig. 53, G-I, K). FEMALE: Medium-sized, elongate-conical, 1.56-1.75 times as long as wide; thin, translucent, almost transparent. Spire conical, slightly convex in outline, about 1.25-1.55 times length of aperture. Umbilicus open, narrow, about one-tenth to one-twelfth diameter of shell. Perforation partially obscured by columellar lip but almost always open. Periostracum tan or light grayish brown, with a dull luster. Sculpture consisting of numerous fine, usually distinct, close incremental striations which are of nearly equal intensity over surface of the shell. 4.5-4.9 whorls. Nuclear whorl 0.24-0.26 mm in diameter, grayish white, protruding. Remaining whorls moderately arched, weakly shouldered. Last whorl descending slightly to the aperture. Suture deeply impressed. Aperture elliptical, occasionally ovate; about 0.71-0.81 times as wide as long. Plane of aperture at about 10° to the axis of the shell. Peristome complete across parietal margin at maturity; usually adnate to preceding whorl, but may be slightly separated. Outer lip thin, sharp, slightly sinuous in lateral profile. Columellar margin very slightly reflected, appressed over umbilical region obscuring perforation.

Length, 2.60-2.95 mm; diameter, 1.55-1.80 mm; height of aperture, 1.05-1.30 mm; length of aperture, 1.12-1.40 mm; width of aperture, 0.82-1.05 mm; length/aperture, 2.27-2.54; 1.48-1.75 whorls/mm.

Occasional specimens may be as long as 3.8 mm and have as many as 5.5 whorls.

Measurements of type: length, 2.90 mm; diameter, 1.75 mm; height of aperture, 1.20 mm; length of aperture, 1.40 mm; width of aperture, 1.02 mm; 4.7 whorls; nuclear whorl, 0.26 mm in diameter.

FIGURE 13

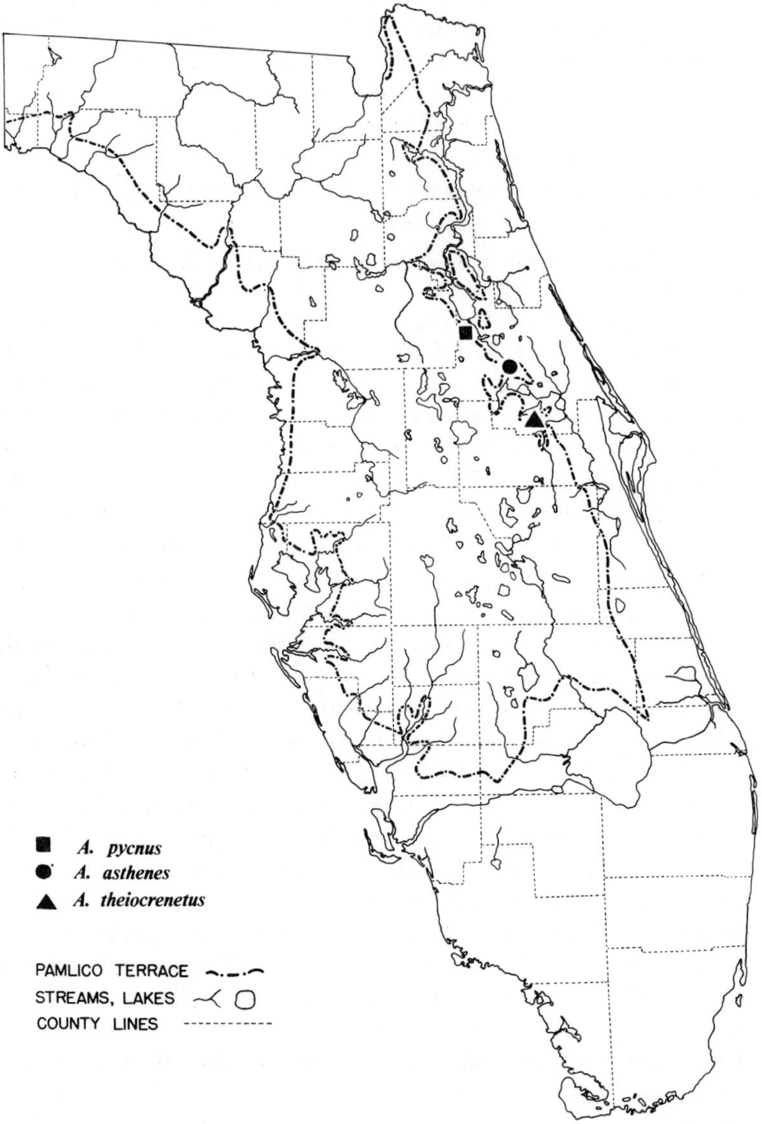

The geographic distribution of *Aphaostracon pycnus* new species, *Aphaostracon asthenes* new species, and *Aphaostracon theiocrenetus* new species.

MALE: The male is shorter and stouter than the female and has only 3.3-3.6 whorls. The aperture is almost always separated from the preceding whorl, and is broadly ovate or nearly triangular in shape due to the nearly straight parietal margin. The aperture is about 0.76-0.85 times as wide as long.

Length, 1.50-1.70 mm; diameter, 1.08-1.20 mm; height of aperture, 0.70-0.82 mm; length of aperture, 0.80-0.92 mm; width of aperture, 0.65-0.75 mm; length/diameter, 1.33-1.48; length/aperture, 1.95-2.21; 2.06-2.33 whorls/mm.

OPERCULUM (Fig. 30, E). Irregularly ovate, upper left margin concave; chitinous; thin, but not fragile; light grayish-yellow; transparent. Paucispiral, consisting of about 3 rapidly expanding whorls; nucleus located at about one-third of the distance from the left margin to the right margin, and from the base to the apex. Outer surface sculptured with numerous distinct incremental striations; inner surface smooth.

VERGE (Fig. 41, G, H). The verge is a simple elongate structure that does not bear papillae. The basal half of the verge is expanded along its left margin, and is slightly compressed dorsoventrally so that it is broadly elliptical in section. The penis is long and slender and gradually tapers to the tip.

TYPE LOCALITY. Clifton Springs Run, about 2 miles north of Oviedo, Seminole County, Florida. The springs are located along the south shore of Lake Jessup and consist of several small and medium-sized boils that form two pools. The spring run is about 200 yards long and enters the lake along the west side of a fishing camp. Type: *UMMZ* 216510; collected August 22, 1962 by Fred G. Thompson. Paratypes: *UMMZ* 216511 (350), *USNM* 636940 (100), *MCZ* 258011 (85), *ANSP* 301544 (30), *CM* 65-14 (84), *UF* (30).

DISTRIBUTION (Fig. 13). Known only from the type locality.

ECOLOGY. The spring run is about 10-25 feet wide, and varies in depth from a few inches to about 2 feet. It is bound on both sides by high sand banks, and has a soft sand bottom that contains scattered patches of aquatic plants. The water has a strong odor due to a high hydrogen sulfide content. Snails were found only on aquatic vegetation in the spring run, where they were very abundant. No mollusks were found in the immediate vicinity of the boils.

Aphaostracon asthenes new species

SHELL (Fig. 63, C, D; Fig. 53, D-F). FEMALE: Small, conical, 1.45-1.62 times as long as wide; very thin and very fragile. The shell is so delicate that it fractures under the slightest pressure or prodding. Transparent when clean. Spire conical, 1.14-1.40 times height of aperture, straight-sided or slightly convex. Umbilicus rimate. Opening very narrow and obscured by the partially superimposed peristome. Periostracum light gray in color; frequently with a muddy, rusty deposit. Sculpture consisting of very fine, but distinct, continuous incremental striations that are nearly uniform in intensity over surface of shell. 4.1-4.5 whorls. Nuclear whorl rounded, protruding, 0.23-0.25 mm in diameter. Remaining whorls regularly increasing in size, strongly rounded but not shouldered or only slightly so. Suture deeply impressed. Aperture ovate, elliptical, or diamond-shaped; upper corner broadly rounded. Aperture 0.73-0.82 times as wide as long. Plane of aperture nearly parallel to axis of shell. Peristome complete across parietal margin; usually adnate to preceding whorl, occasionally separated. Outer lip thin, sharp, fragile, sinuous in lateral profile. Columellar margin slightly reflected or directed forward. Parietal margin straight or curved.

Length, 2.00-2.42 mm; diameter, 1.31-1.50 mm; height of aperture, 0.87-1.01 mm; length of aperture, 0.91-1.08 mm; width of aperture, 0.72-0.89 mm; length/aperture, 2.14-2.40; 1.86-2.04 whorls/mm.

Measurements of type: length, 2.10 mm; diameter, 1.40 mm; height of aperture, 0.94 mm; length of aperture, 1.03 mm; width of aperture, 0.75 mm; 4.3 whorls; nuclear whorl, 0.25 mm in diameter.

MALE: As is usual in the genus, the male is much smaller than the female, and has about one whorl less. Only one male was collected during the four trips to the type locality, and its shell was so fragile that it was crushed while the verge was being removed. Thus, its measurements are not available.

OPERCULUM (Fig. 30, F). Ovate, apex bluntly pointed; chitinous, thin, fragile; colorless. Paucispiral, consisting of about 3 whorls which rapidly expand in size. Nucleus large, located on about the midline and at about one-third of the distance from the base to the apex. Outer surface sculptured with rather distinct wavy incremental striations. Inner surface smooth.

VERGE (Fig. 41, I). The verge is relatively long and slender. It does not have any papillae along its right margin, nor does it have a conspicuous bulge near the base of its left margin. The vas deferens lies in the usual position along the right margin.

TYPE LOCALITY. Blue Springs, 3 miles west of Orange City, Volusia County, Florida. Type: *UMMZ* 216519; collected April 20, 1962 by Fred G. Thompson. Paratypes: *UMMZ* 216520 (14); same data as the type.

DISTRIBUTION (Fig. 13). Known from the type locality only.

ECOLOGY. Blue Springs is a large spring that gives rise to a wide, deep spring run that flows west for about one-fourth of a mile to the St. Johns River. The spring boil is nearly devoid of rooted vegetation, but the spring run has thick growths of aquatic plants throughout its course. *Aphaostracon asthenes* has been collected only in the spring run where plants and bottom debris were very sparse.

The *pachynotus* Group

This group consists of a heterogenous assemblage of three species that are placed together for convenience. The group is characterized by 3 or more papillae on the right margin of the verge. Two species, *A. pachynotus* and *A. xynoelictus,* consistently have 3 papillae. The third species, *A. chalarogyrus,* has 4-6 papillae. *Aphaostracon pachynotus* is generally distributed from the Miami region north to Sanford. The other two species are confined to single springs in the north central part of the peninsula. This disjunct distribution suggests that they are not commonly related within the genus, but probably have been derived independently from the *rhadinus* group. The three species may be distinguished as follows.

A. pachynotus—The shell is cylindric-conical in shape, and is solid and opaque. It has 4.5-4.9 moderately rounded whorls, with a moderately impressed suture. The aperture is ovate and is usually adnate to the preceding whorl.

A. xynoelictus—The shell is conical in shape, and is thin and subtransparent. It has 4.0-4.5 strongly rounded whorls that are not shouldered. The suture is deeply impressed. The ovate or elliptical aperture is loosely appressed against or free from the preceding whorl.

A. chalarogyrus—The shell is conical in shape, and is thin and transparent. There are 4.2-4.5 strongly arched, strongly shouldered whorls with a very deeply impressed suture. The aperture is relatively broad, and is widely separated from the preceding whorl.

Aphaostracon pachynotus new species

SHELL (Fig. 64, A-C; Fig. 54, A-C, M). FEMALE: Cylindric-conical, 1.45-1.77 times as long as wide. Shell moderately thick, solid; opaque, occasionally translucent. Spire conical, straight-sided or slightly convex, 1.61-1.65 times the height of the aperture. Umbilicus open, but obscured by the partially superimposed peristome. Opening about one-tenth to one-twelfth the diameter of shell. Periostracum usually rusty brown or brownish gray, usually free of mineral deposits. Sculpture consisting of very fine, indistinct, discontinuous incremental striations. 4.5-4.9 whorls. Nuclear whorl 0.20-0.25 mm in diameter, protruding, grayish-white in color. Remaining whorls weakly or moderately arched, *never* shouldered, regularly increasing in size. Suture moderately impressed. Aperture ovate; upper corner angulate. Plane of aperture nearly parallel to axis of shell. Inside of aperture livid white at maturity. Aperture 0.68-0.81 times as wide as long. Peristome complete across parietal margin at maturity; adnate to preceding whorl. Outer lip thick but sharp, very slightly arched forward in lateral profile or straight-edged. Columellar margin thickened, directed forward or only very slightly reflected.

Length, 2.10-2.85 mm; diameter, 1.30-1.70 mm; height of aperture, 0.85-1.20 mm; length of aperture, 0.97-1.25 mm; width of aperture, 0.75-0.91 mm; length/aperture, 2.16-2.65; 1.72-2.09 whorls/mm.

Measurements of type: length, 2.55 mm; diameter, 1.50 mm; height of aperture, 1.10 mm; length of aperture, 1.18 mm; width of aperture, 0.82 mm; 4.7 whorls; nuclear whorl 0.21 mm in diameter.

This species shows considerable geographic variation. Specimens from the northern part of the range tend to be large and stocky, while some specimens from southern areas are relatively small and slender. Occasional specimens from the Miami region are very attenuate and have a much reduced aperture, and some specimens from Dade and Broward Counties are atypical in that the peristome may be slightly separated from the preceding whorl and the upper corner of the aperture may be rounded. Most southern population samples do not show these variations, and when they do occur, they seldom involve more than one-fourth of the population.

MALE: Males differ from females in that they are about 1 mm shorter, are stockier, they have about 1 whorl less, and they have a broader, more rounded aperture. Males were collected on only three occasions, and on two of these only a single specimen was found. A large series of males was collected from Lake Okeechobee, at Clewiston, Hendry County. This series is atypical in that specimens of both sexes are thinner-shelled and grayer than is normal. They are normal in other characters, and measurements of the males are probably representative of the species.

Length, 1.25-1.49 mm; diameter, 0.96-1.10 mm; height of aperture, 0.60-0.72 mm; length of aperture, 0.65-0.75 mm; width of aperture, 0.50-0.58 mm; 3.4-3.7 whorls; length/diameter, 1.24-1.35; length/aperture, 1.98-2.12; 2.48-2.80 whorls/mm; aperture 0.77-0.82 times as wide as long.

OPERCULUM (Fig. 30, G). Oval, apex pointed or bluntly angulate, thin, chitinous, transparent, colorless, fragile. Paucispiral, consisting of about 3 rapidly expanding whorls; nucleus located at about one-third of the distance from the left margin to the right and from the base to the apex. Outer surface sculptured with a few very fine incremental striations. Smooth inner surface.

VERGE (Fig. 41, J). The verge bears 3 conical papillae along the right margin, which are uniform in size and shape. The left margin bears a large lateral bulge near the base. The penis is long and slender and curves up and to the left.

TYPE LOCALITY. Canal along Florida State Highway 706, 7.4 miles west of Jupiter, Palm Beach County, Florida. Type: *UMMZ* 216521; collected March 23, 1962 by Fred G. Thompson. Paratypes: *UMMZ* 216522 (72), *MCZ* 258010 (5), *ANSP* 301540 (5), *USNM* 636943 (10), *UF* (5); same locality as the type.

DISTRIBUTION (Fig. 14). Known only from eastern Florida, where it has been found from the upper half of the St. Johns River near Sanford, south to the Miami region.

LOCALITIES. Florida: *Broward Co.*—canal along U.S. Hwy. 27, 2.5 mi. n. Dade Co. line; canal along U.S. Hwy. 27, 2.7 mi. n. Dade Co. line; canal along U.S. Hwy. 27, 6.3 mi. n. Dade Co. line. *Dade Co.*—Snake Creek Canal #9; numerous places in canal along U.S. Hwy. 441. *Glades Co.*—canal 11 mi. e. LaBelle; canal 4.4 mi. e.

FIGURE 14

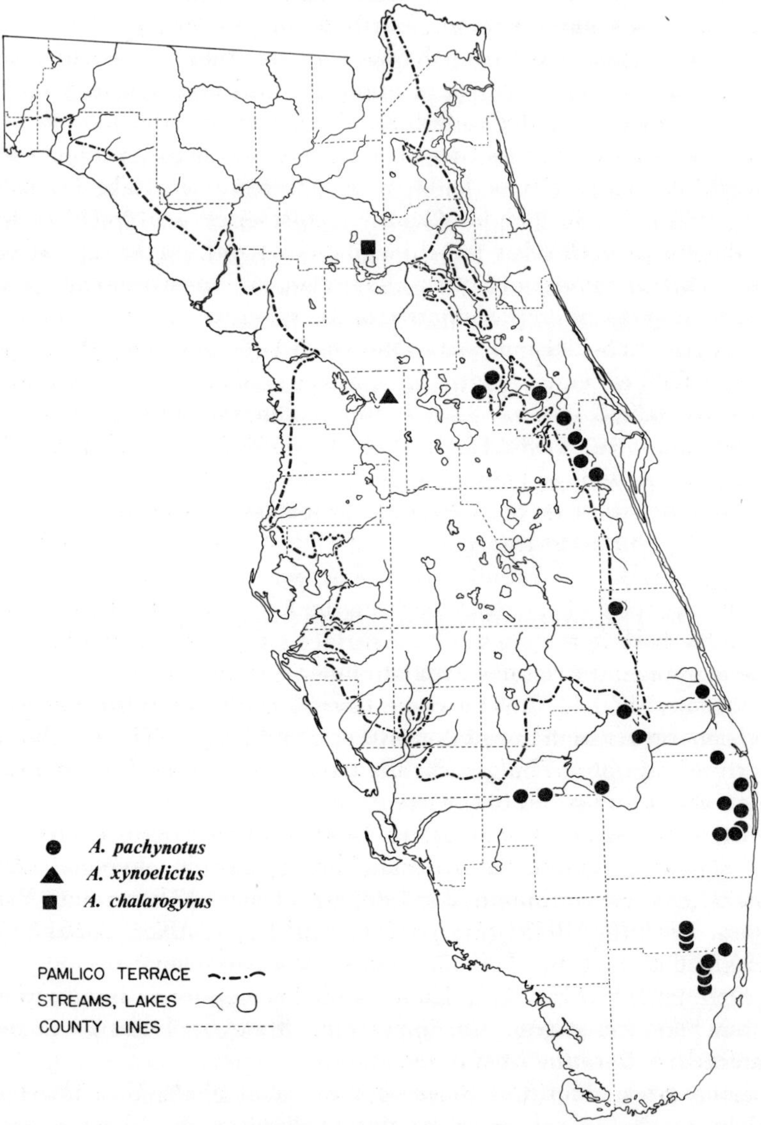

The geographic distribution of *Aphaostracon pachynotus* new species, *Aphaostracon xynoelictus* new species, and *Aphaostracon chalarogyrus* new species.

LaBelle. *Hendry Co.*—Lake Okeechobee, Clewiston. *Indian River Co.*—Blue Cypress Lake. *Lake Co.*—Seminole Springs, 3.5 mi. n.e. Sorrento; Blackwater Creek, 2 mi. s.w. Cassia. *Okeechobee Co.*—Lake Okeechobee, 4.5 mi. s.e. Okeechobee. *Orange Co.*—St. Johns R., 9 mi. w. Cocoa; St. Johns R., 4 mi. e. Christmas; small creek 2.1 mi. e. Christmas; small creek 12.1 mi. s.e. Bithlo. *Palm Beach Co.*—Lake Okeechobee, 2.4 mi. w.n.w. Lake Harbor; canal 4.6 mi. e.n.e. Loxahatchee; canal, Loxahatchee; canal 5 mi. w. Delray Beach; canal 3.8 mi. w.n.w. Delray Beach; canal 4 mi. w.s.w. Lake Worth; canal 3.5 mi. n.w. West Palm Beach; canal 4.7 mi. w. Greenacre City. *Seminole Co.*—creek 7.4 mi. s.e. Geneve. *St. Lucie Co.*—creek 3 mi. s. Port St. Lucie. *Volusia Co.*–St. Johns R., 4 mi. e. Sanford.

Ecology. This snail is generally distributed in canals, lakes, and streams, where it may be found on plants, debris, or bottom sediments. It is tolerant of a variety of conditions, and seems to be equally abundant in clear or turbid water, and on muck as well as sand bottoms. Aquatic vegetation may be absent at stations where it is found, although it was generally collected from mats of hyacinths and other angiosperms.

Aphaostracon xynoelictus **new species**

Shell (Fig. 64, D, E; Fig. 54, D-F, L). female: Small, conical, about 1.37-1.61 times as long as wide; thin, translucent, occasionally transparent. Spire conical, 1.08-1.50 times height of aperture, straight-sided or slightly convex. Umbilical perforation open, about one-eighth to one-tenth the diameter of the shell. The peristome passes partially over the area, but does not obscure it to any significant extent. Periostracum dull gray or olive-gray, frequently colored with a dull rusty deposit. Clean shells with a glossy luster. Sculpture consisting of very fine but distinct, continuous, incremental striations. 4.0-4.5 whorls. Nuclear whorl strongly protruding, 0.22-0.25 mm in diameter. Remaining whorls regularly increasing in size, strongly arched, rarely shouldered. Suture deeply impressed. Aperture ovate or elliptical, upper corner rounded. Aperture about 0.71-0.79 times as wide as long. Plane of aperture nearly parallel to axis of shell. Peristome complete across parietal region, appressed against or slightly free from preceding whorl (as in type), occasionally more separated. Outer lip thin, fragile, sinuous in lateral profile. Columellar margin directed forward, only slightly reflected at the most. Parietal margin is slightly curved.

Length, 2.00-2.45 mm; diameter, 1.35-1.60 mm; height of aperture, 0.89-1.05 mm; length of aperture, 0.94-1.13 mm; width of aperture, 0.72-0.84 mm; length/aperture, 2.08-2.50; 1 .75-2.05 whorls/mm.

Measurements of type: length, 2.45 mm; diameter, 1.55 mm; height of aperture, 1.01 mm; length of aperture, 1.10 mm; width of aperture, 0.79 mm; 4.5 whorls; diameter of nuclear whorl, 0.23 mm.

MALE: Males differ from females principally in being smaller, in having a more broadly ovate aperture, and in having the peristome more widely separated from the preceding whorl. Measurements for the males are: length, 1.60-1.75 mm; diameter, 1.13-1.25 mm; height of aperture, 0.77-0.84 mm; length of aperture, 0.79-0.89 mm; width of aperture, 0.65-0.72 mm; 3.5-3.6 whorls; length/diameter, 1.38-1.46; length/aperture, 2.05-2.14; 2.06-2.18 whorls/mm; aperture 0.81-0.83 times as wide as long.

OPERCULUM (Fig. 30, H). Broadly ovate, apex bluntly pointed or rounded, chitinous, thin, transparent, colorless or grayish. Paucispiral, consisting of about 2.75-3.00 large, rapidly expanding whorls. Nucleus located at about one-third of the distance from the left margin to the right and from the base to the apex. Outer surface densely sculptured with fine incremental striations along the sutures. Smooth inner surface.

VERGE (Fig. 41, K). The verge is a small stocky structure. Its right margin bears 3 relatively large, conical papillae below the middle, and its left margin bears a large lateral bulge near the base. The penis is relatively stout, and rapidly tapers to a point.

TYPE LOCALITY. Fenney Springs, 2 miles east of Coleman, Sumter County, Florida. Type: *UMMZ* 216517; collected August 23, 1962 by Fred G. Thompson. Paratypes: *UMMZ* 216518 (20); same data as the type.

DISTRIBUTION (Fig. 14). Known only from the type locality.

ECOLOGY. Fenney Springs consists of two moderately large, vertical-sided pools that were formed by a collapsed limestone cavern. The pools drain into a rather large spring run that flows west for several miles to Lake Panasoffkee. The spring pools were devoid of macroscopic vegetation at the time that I was there, but the spring run had several scattered patches of aquatic angiosperms and *Chara*. *Aphaostracon xynoelictus* was found only in the spring pools and in the spring run in the immediate vicinity of the springs. Although it was sparse, it was generally distributed on the substrate, debris, and vegetation. It was associated with *Cincinnatia floridana,* which was abundant throughout the spring run.

Aphaostracon chalarogyrus new species

SHELL (Fig. 64, F, G; Fig. 54, G-K). FEMALE: Medium-sized, occasionally large; conical, about 1.36-1.59 times as long as wide. Thin, fragile, transparent. Spire conical, 1.22-1.70 times the length of the aperture, slightly convex in outline, occasionally straight-sided. Umbilical perforation narrow, less than one-twelfth the diameter of shell, partially shielded by peristome. Periostracum light gray or gray-brown with a dull luster. Sculpture consisting of very fine, indistinct incremental striations, which frequently are supplemented by distinct, evenly spaced, sharp axial threads which are most prominent on the earlier whorls and in the umbilical region. 4.2-4.5 whorls. Nuclear whorl 0.23-0.25 mm in diameter, protruding. Remaining whorls strongly arched, distinctly shouldered, and tending to be flattened below the periphery. In many specimens most of the last whorl may be separated from the preceding whorl. This and the very deep suture along the earlier whorls give the shell the appearance of being loosely coiled. Aperture broadly ovate or rhomboid, 0.78-0.90 times as wide as long. Plane of aperture nearly parallel to axis of shell. Peristome complete across parietal region, free from preceding whorl by a considerable distance. Outer lip thin, sharp, sinuous in lateral profile. Columellar margin thin, fragile, slightly reflected. Parietal margin straight or nearly so.

Length, 2.20-2.60 mm; diameter, 1.45-1.80 mm; height of aperture, 0.87-1.17 mm; length of aperture, 0.94-1.20 mm; width of aperture, 0.82-0.94 mm; length/aperture, 2.22-2.70; 1.69-1.95 whorls/mm.

Two generations of female shells are common. The younger generation (Fig. 54, H, I), upon which the description and measurements are based, is dominant and corresponds to the size range of the other species of *Aphaostracon* discussed in this paper. The older generation (Fig. 54, G) consists of a few individuals which reach a much larger size. These large individuals tend to be scalariform due to the partial separation of the third and fourth whorls, and there is a tendency for the aperture to be more closely associated with the preceding whorl. The type specimen is selected from these larger individuals because it demonstrates the loosely coiled scalariform nature of the whorls better than do smaller specimens. Measurements of large female shells are as follows: length, 3.00-4.00 mm; diameter, 1.90-2.60 mm; height of aperture, 1.12-1.49 mm; length of aperture, 1.22-1.49 mm; width of aperture, 0.94-1.34 mm; 4.9-5.5 whorls; length/diameter, 1.48-1.82; length/aperture, 2.48-2.86; 1.37-1.59 whorls/mm. The lower extremes of these measurements are for the type.

MALE: Males are like females in most features, except that they are much smaller. Measurements of males are: length, 1.53-1.82 mm; diameter, 1.20-1.40 mm; height of aperture, 0.77-0.94 mm; length of aperture, 0.79-0.97 mm; width of aperture, 0.70-0.82 mm; 3.4-3.9 whorls; length/diameter, 1.22-1.42; length/aperture, 1.94-2.17; aperture 0.83-0.92 times as wide as long; 2.06-2.22 whorls/mm.

OPERCULUM (Fig. 30, I). Rotundate, thin, fragile, transparent, light gray in color. Paucispiral, consisting of about three and one-quarter whorls, the last of which rapidly expands in size and comprises most of the operculum; nucleus located at about one-third of the distance from the left margin to the right and from the base to the apex. Outer surface sculptured with a few faint incremental striations.

VERGE (Fig. 41, L, M). The verge is a stout sickle-shaped structure that is dorsoventrally compressed. Its right margin bears 4-6 conical papillae that are nearly uniform in size. The right margin bears a large lateral expansion near the base. The penis is stout and tapers to a fine point.

TYPE LOCALITY. Magnesia Springs, 3.7 miles west of Hawthorne, Alachua County, Florida. Type: *UMMZ* 216515: collected March 19, 1963 by Fred G. Thompson. Paratypes: *UMMZ* 216516 (400), *USNM* 636942 (100), *MCZ* 258007 (100), *ANSP* 301547 (100), *CM* 65-13 (92), *UF;* same data as the type.

DISTRIBUTION (Fig. 14). Restricted to the type locality.

ECOLOGY. The spring has been impounded by a large circular concrete wall and acts as a water source for a public swimming pool. The impounded spring pool is about 10 feet deep; it has a fine calcareous silty sand bottom which supports thick patches and mats of filamentous algae. Adjacent to the spring is a shallow rectangular cement overflow pool that is about 6 feet wide by about 10 feet long and only a few inches deep. The bottom of the pool has a fine calcareous ooze overlying the cement, and the pool usually has large mats of *Spirogyra* floating on the surface. What remains of the spring run drains into a flatwood swamp that borders a small, sand-bottomed creek.

Aphaostracon chalarogyrus was abundant in the drainage pool by the spring, where it was generally found on the algal mats and in the ooze. It was less common in the spring pool, although it was generally distributed over the cement wall, bottom silt, and vegetation. The spring run, the flatwood swamp, and the creek were devoid of hydrobiid snails, and the only other aquatic mollusk observed there was *Pomacea paludosa*.

THE *SOMATOGYRUS* TRIBE

The six genera known to comprise this tribe are distinguished as follows.

Somatogyrus. Shell small or medium-sized, conical to subglobose, thick, white or light gray, usually imperforate; suture weakly or moderately impressed; peristome usually incomplete across parietal wall; body whorl large, not descending to the aperture; columellar margin of aperture reflected as a wide, thick, white callus. Central tooth of radula with 4 basocones on each side. The anatomical accounts that have been published on this genus are too superficial to allow comparison of the verge with that of other genera. Material that I have examined indicates that the genus as presently conceived is at least biphyletic.

Birgella. Shell very large (10 mm), subglobose, comparatively thin, fragile, light horn colored, narrowly perforate; suture moderately impressed; peristome complete across parietal wall as a thin transparent deposit; body whorl large, ample, not descending; columellar margin of aperture thin, weakly reflected. Central tooth with 3 basocones on each side. Accessory lobe of verge slender; no dermal glands have been reported on the verge (F. C. Baker, 1928: 156-57; Berry, 1943: 49-50).

Notogillia. Shell medium-sized or large, thick, ovate or globose, dark yellow to olivaceous, imperforate; suture moderately impressed; body whorl globose, rapidly descending to aperture along last half whorl; peristome complete; upper corner of aperture very thick; columellar margin very thick, but not reflected. Central tooth with 1 or rarely 2 basocones on each side. Verge with a characteristic pattern of glandular crests as described below.

Spilochlamys (new genus). Shell medium-sized, moderately thick, conical, olivaceous, openly umbilicate; umbilicus may be circumscribed by a thickened ridge or keel; suture deeply impressed; body whorl not conspicuously enlarged, regularly descending to the aperture with rest of whorls; peristome complete; not thickened, except along outer lip in some instances. Central tooth with a single basocone on each side. Mantle richly marked with large reddish-brown blotches and spots. Verge with characteristic arrangement of glandular crests as described below.

Cincinnatia. Shell small or medium-sized, thin or only moderately thick, conical or cylindric-conical, olivaceous, usually rimate or imperforate, but may be umbilicate; suture variable; peristome complete across parietal wall; body whorl not conspicuously enlarged, regularly descending; aperture simple, columellar margin thin, weakly reflected. Central tooth with a single basocone on each side. Verge with a characteristic arrangement of glandular crests as described below.

Nymphophilus. Shell medium to large, conical-trochoid, carinate, white or light tan, narrowly perforate; suture very shallow, sides of spire nearly straight. Peristome incomplete across parietal margin; whorls regularly increasing in size, not descending at the aperture; columellar margin weakly reflected and very slightly thickened. Radula unknown. Verge with a lateral blade-like penis and a large terminal lobe. A single large gland on tip of terminal lobe, and 1 or more smaller glands scattered over dorsal surface of verge. Endemic to the Cuatro Cienegas Basin of Mexico (Taylor, 1966).

Notogillia, Spilochlamys, and *Cincinnatia* occur in peninsular Florida, and *Somatogyrus* may occur in the panhandle of the state.

Notogillia Pilsbry, 1953

Notogillia Pilsbry, 1953, Monogr. Acad. Nat. Sci. Phila., 8: 440.
Type Species: *Hydrobia wetherbyi* Dall, 1885 (designated by Pilsbry, 1953).

This monotypic genus is confined to Florida, and adjacent regions of Georgia and Alabama. It is characterized by a large, globose, thick, opaque shell that is imperforate, and has a large, globose body whorl that descends to the aperture very rapidly along the last half of the whorl. The aperture is ovate or subcircular, and has a very thick columellar margin and a very thick upper corner where the outer lip joins the body whorl. The periostracum is straw yellow to dark olive-brown, and has distinct incremental striations that become more intense near the aperture.

The central tooth of the radula has a single basocone on each side. The mantle lacks pigmentation, or may be moderately shaded with melanophores. The tentacles are short and stocky. The verge has a blade-like penis along the right margin, and a heavy mid-ventral ridge which bears 8-11 narrow transverse glands. The glands on the outer surface of the verge are arranged in a characteristic pattern, which is discussed below.

Notogillia is probably a relict genus that has existed in Florida since at least the Pliocene. Pilsbry (1953, 441) described a species from the Pliocene of St. Petersburg, which is not very different from the present form.

Notogillia wetherbyi is not apt to be confused with any other hydrobiid snail in Florida, for its large size, thick shell, dark color, and characteristic aperture distinguish it from all other species that are known to occur within its range, and indeed, within the United States.

Notogillia wetherbyi (Dall)

Hydrobia wetherbyi Dall, 1885, Proc. U.S. Natl. Museum, 8 (17): 258; pl. 17, fig. 10.

Gillia wetherbyi (Dall) Walker, 1918, Misc. Publ. Museum Zool. Univ. Mich., (6): 146.

Notogillia wetherbyi (Dall) Pilsbry, 1953, Monogr. Acad. Nat. Sci. Phila., 8: 440; pl. 65, figs. 6, 9; text fig. 1a.—Clench and Turner, 1956, Bull. Florida State Museum, 1 (3): 124-25; pl. 7, fig. 6.

SHELL (Fig. 65, D, E). Stout, thick, opaque; ovate to broadly conical; spire usually shorter than the length of the aperture (highly variable). Imperforate, occasionally very narrowly rimate with the umbilicus reduced to a groove behind the lip. Straw yellow to dark olive-brown in color, usually with a dull luster; edge of peristome dark brown or black in mature specimens; interior of aperture white. Sculpture consisting of distinct incremental striations that usually become more intense near the aperture; very fine spiral sculpture may occasionally be evident. 4.2-5.2 whorls; nuclear whorl usually protruding, 0.25-0.35 mm in diameter; early whorls of spire usually depressed, but rarely flattened; body whorl globose, becoming basal in position. Suture deeply impressed, rapidly descending to aperture along last half of body whorl. Aperture ovate or subcircular; upper corner obtusely angulate. Peristome complete; outer lip thick, almost blunt-edged, straight in lateral profile; upper corner of peristome and columellar margin conspicuously thickened; parietal margin between these two areas noticeably thickened, but to a lesser extent.

Length of shell, 4.4-7.7 mm; diameter of shell, 3.9-5.8 mm; length of aperture, 2.5-4.0 mm; width of aperture, 2.1-3.4 mm; length/diameter, 1.13-1.57; length/aperture, 1.62-2.40; whorls/length, 0.61-0.96.

Measurements of the lectotype: length of shell, 7.1 mm; diameter of shell, 5.0 mm; length of aperture, 3.1 mm; width of aperture, 3.1 mm; 5.0 whorls; nuclear whorl 0.35 mm in diameter, eroded.

Most populations tend to have their own shell characteristic. Size is highly variable (Table 4). Specimens from lakes tend to be larger than specimens from rivers and large springs. Specimens from small springs and spring runs frequently are dwarfed. The relative length of the spire also is variable, but is independent of the habitat occupied by the snail.

Operculum (Fig. 31, A). Broadly oval, consisting of about two and one-quarter rapidly expanding whorls; nucleus located at about one-third of the distance from the columellar margin to the outer margin, and from the base to the apex; outer margin and basal margin uniformly rounded; parietal margin straight or very slightly concave. Outer surface coarsely and densely striate; inner surface smooth.

Animal (Fig. 26, A, B). The foot is broad, bluntly rounded posteriorly, indented anteriorly, with the two anterior corners extending slightly to the sides. The snout is broad and stout, and the tentacles are relatively short and thick. When extended the tentacles do not exceed two-thirds of the length of the snout. The animal is only slightly pigmented, and most populations exhibit the same color patterns. The muzzle is white, and the sides and top of the snout are encircled by about 15-20 narrow, diffuse, gray bands. The bands are about as wide as their intervals, and are discontinuous ventrally. The pedicel and the top of the foot have a few dark flecks that are uniformly scattered over their surfaces, giving these parts

TABLE 4.—MEASUREMENTS AND RATIOS OF SHELLS OF NINE POPULATION SAMPLES OF *Notogillia wetherbyi* SHOWING INTRASPECIFIC VARIATION. TEN MATURE SPECIMENS WERE SELECTED AT RANDOM FROM EACH POPULATION SAMPLE. ALL MEASUREMENTS ARE IN MILLIMETERS

Locality	Length	Diameter	Whorls	Length / Diameter	Length / Aperture	Whorls / Length
Weeki Wachee Springs	5.3 - 6.0	4.3 - 4.9	4.4 - 5.0	1.20 - 1.29	1.64 - 1.90	0.76 - 0.86
Crystal Springs	5.5 - 7.4	3.9 - 5.3	4.2 - 4.9	1.39 - 1.56	2.11 - 2.32	0.66 - 0.76
Rock Springs	4.4 - 5.5	3.9 - 4.6	4.2 - 4.7	1.13 - 1.28	1.63 - 1.89	0.82 - 0.96
Lake Apopka	5.4 - 6.3	4.5 - 4.8	4.3 - 4.8	1.17 - 1.34	1.67 - 1.90	0.74 - 0.84
Lake Eustis	6.4 - 7.7	4.9 - 5.8	4.7 - 5.1	1.19 - 1.36	1.62 - 2.00	0.61 - 0.76
Alexander Springs	6.2 - 6.9	4.8 - 5.3	4.6 - 5.0	1.22 - 1.40	1.75 - 2.06	0.70 - 0.80
Silver Springs	5.4 - 6.4	4.1 - 4.8	4.5 - 5.1	1.21 - 1.39	1.84 - 2.10	0.78 - 0.87
Rainbow Springs	6.5 - 7.3	5.0 - 5.7	4.8 - 5.2	1.20 - 1.38	1.75 - 2.15	0.71 - 0.75
Ichetucknee Springs	6.4 - 7.4	4.2 - 4.8	4.4 - 5.1	1.42 - 1.57	2.09 - 2.40	0.65 - 0.70

a very light gray color. The dorsal surface of the body has the same gray color and two darker gray bands that begin between the tentacle and extend posteriorly under the mantle. The dorsal surfaces of the tentacles bear two gray blotches, one near the base above the eye, and the other over the distal third of the tentacle. The mantle is usually unpigmented. Occasional specimens may have a diffuse, mottled gray pattern over the mantle, and some specimens may be even darker.

VERGE (Fig. 42, E-I). The verge is a broad ponderous structure. It originates on the right side of the nape with its left margin lying on the middorsal line. When relaxed it is folded back upon itself with its terminal end pointing forward and slightly to the right. When erect it extends forward and to the right. The right side of the verge bears a flat, lanceolate, blade-like penis that originates near the middle. A simple tubular vas deferens passes through the verge and the penis close to their right margins. The outer surface of the accessory lobe of the verge bears no conspicuous enlargements, except for a gentle bulge along the right margin distal to the origin of the penis. A large midventral ridge lies on the distal half of the verge, causing the verge to be triangular in cross section in this region.

Four glands on the outer surface of the verge are consistently present (Fig. 42, E, H) although they are variable in shape and size. Near the right margin, about halfway between the base and the origin of the penis, is a large circular gland that is raised on a short pedicel. About halfway between this gland and the penis, and close to the right margin, is another gland that is long and narrow, and extends onto the dorsal surface of the penis for one-half or more of its length. Another long narrow gland lies parallel to the edge of the bulge along the right margin of the accessory lobe. On the opposite edge of the accessory lobe is a large apical crest, which is visible on both surfaces of the verge. Scattered over the dorsal surface in the area between these four glands are numerous small narrow glands that generally lie parallel to the axis of the verge.

On the inner surface of the verge, lying along the midventral crest is a series of 8-11 narrow, transverse glands that may coalesce or break up at random (Fig. 42, E, I). At the end of this series and near the base is a larger, narrow gland that is variable in size and shape, and may appear to be continuous with the series on the crest. An additional long narrow gland lies along the midventral surface of the penis. Along the opposite margin of the ventral surface are two large glands. One is the apical crest, discussed above. The other gland lies below the apical crest, and extends in from the margin.

RADULA (Fig. 35, C). Ten specimens examined. The ribbon contains 59-65 transverse rows of teeth. The central tooth is about 40μ wide. It has a conspicuous basolateral extension on each side, and the center of the basal margin has a narrow tongue-like projection. The basolateral projections have a reinforced crest on their faces. A single basocone is located on each side of the tooth at the top of the crests. The reflection has a large lanceolate mesocone, and 4-6 ectocones on each side. The lateral tooth is about 100μ long, and is hatchet-shaped, with a long lateral shaft. The reflection bears a large mesocone, 1 or 2 entocones, and 4 or 5 ectocones. The inner marginal tooth has about 20 cusps on the reflection. The shaft bears a long, wide flange-like extension. The outer marginal tooth has about 30 minute cusps, which are too small to be clearly distinguished with oil immersion. The base of the shaft has a conspicuous flange-like extension along its outer margin.

TYPE LOCALITY. Lake Eustis, Lake County, Florida. Lectotype: *USNM* 32123, specimen figured by Dall.

DISTRIBUTION (Fig. 15). This species is common and widespread throughout the northern half of the peninsula west of the St. Johns River and as far north as the Suwannee River System. It is also known from scattered localities in Jackson County and Calhoun County in the Florida panhandle, and from Decatur County and Seminole County, Georgia (Clench and Turner, 1956: 125). Hubricht (1963, 152) has recorded it from Barbour County, Alabama.

LOCALITIES. Florida: *Alachua Co.*—Santa Fe R., 3 mi. n.w. High Springs. *Citrus Co.*—Withlacoochee R., Dunnellon; Withlacoochee R. below dam at reservoir; Withlacoochee R., 4 mi. above reservoir dam. *Dixie Co.*—spring 2 mi. s. Rock Bluff Ferry at Fla. 340. *Gilchrist Co.*—Bell Springs, 0.8 mi. n.w. Fanning Springs; Sun Springs, 3 mi. w. Curtis. *Hernando Co.*—Weeki Wachee Springs. *Lafayette Co.*—Troy Springs about 6 mi. n.w. Branford. *Lake Co.*—Lake Apopka, Montverde; Lake Eustis, Eustis; Lake Eustis, ½ mi. w. Eustis; Lake Eustis, Grand Island; Lake Yale, 3 mi. n.w. Grand Island; s.e. end Lake Harris at Fla. 19; Lake Harris, 2 mi. s. Leesburg; Blue Springs, 1 mi. w. Yalaha; Echo Glen Springs at Yalaha; Alexander Springs; Alexander Springs Run at Fla. 445; Seminole

FIGURE 15

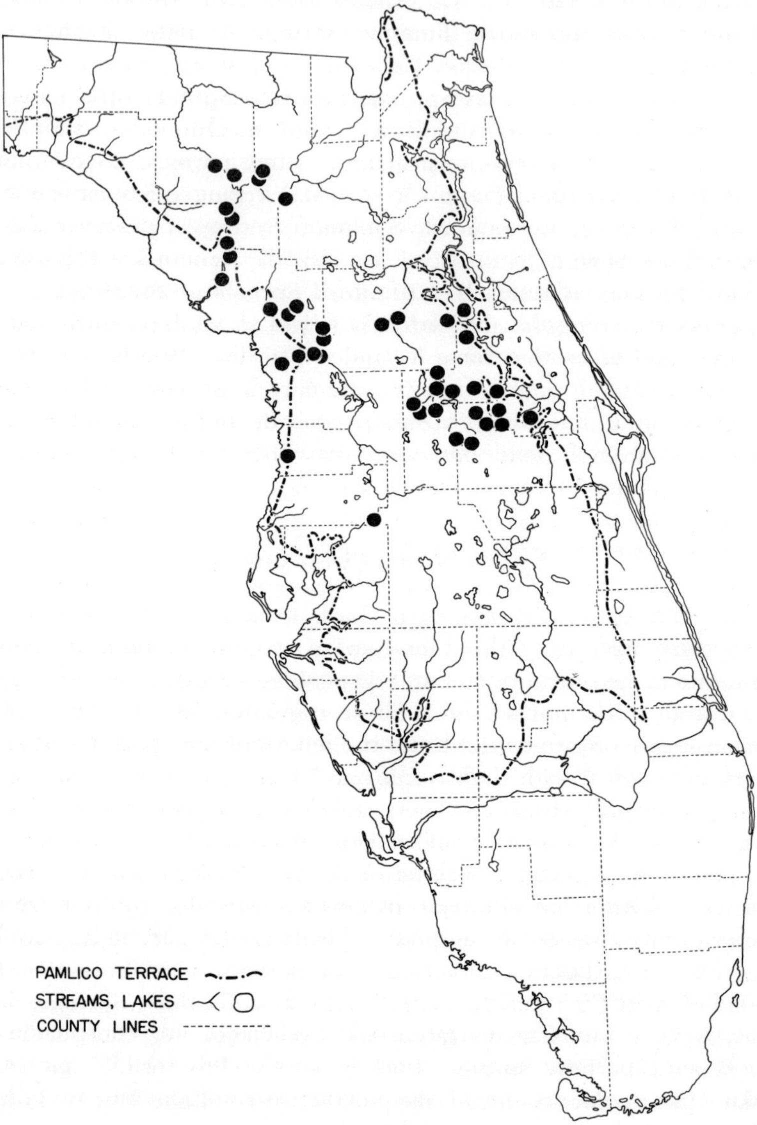

The geographic distribution of *Notogillia wetherbyi* (Dall) in the Florida peninsula.

Springs, 3.5 mi. n.e. Sorrento. *Levy Co.*—creek at Gulf Hammock; Wekiva Springs Run, 1.1 mi. e. Gulf Hammock; Wekiva Springs, 6.3 mi. e. Gulf Hammock; Fanning Springs; Manatee Springs, 6.0 mi. w. Hardeetown. *Marion Co.*—Juniper Springs; Oklawaha R., 8.5 mi. e.n.e. Ocala; Silver Springs; Rainbow Springs Run, 1 mi. e. Dunnellon; Rainbow Springs Run, 2.8 mi. n. Dunnellon; Rainbow Springs. *Orange Co.*—Wekiva Springs; Rock Springs; Lake Apopka at Winter Garden. *Pasco Co.*—Crystal Springs. *Seminole Co.*—Wekiva R., 9 mi. w. Sanford; Sanlando Springs. *Suwannee Co.*—Ichetucknee Springs Run, 1 mi. n. U.S. 27; Ichetucknee Springs; Telford Springs at Luraville; Branford Springs at Branford.

ECOLOGY. *Notogillia wetherbyi* is found in springs, spring runs, streams, and lakes that have a sand substrate. Usually specimens are very abundant where there is some current or wave action. They are generally found buried in sand that is devoid of vegetation, but occasional animals may be found crawling on aquatic plants in streams with a moderate or fast current.

Spilochlamys new genus

The shell is conical or ovate; thick, transparent or translucent; olivaceous, with or without fine spiral stripes. Umbilicus openly perforate or rimate; a circumumbilical ridge or keel is usually present due to thickening of the shell or angulation of the body whorl in that region. Spire 0.6-1.1 times the length of the aperture, usually short and conical with slightly convex sides. Sculpture consisting of fine incremental striations; very seldom are spiral striations detectable. 4.1-5.4 whorls; nuclear whorl protruding, 0.25-0.30 mm in diameter. Remaining whorls strongly shouldered; suture deeply impressed. Aperture elliptical, ovate or auriculate; interior white; upper corner rounded or angulate; lower corner parabolic, may be angulate. Peristome complete across parietal margin; outer lip rounded, with or without an internal callus, straight-edged in lateral view; columellar margin weakly reflected, not conspicuously thickened; parietal margin thin to moderately thick; parietal-columellar angle making a conspicuous indentation into umbilical region.

The operculum is ovate, transparent, chitinous, amber colored, and paucispiral. The nucleus is located about one-third of the distance from the columellar margin to the outer margin, and about one-fourth of the distance from the base to the apex. The operculum consists of about 2.2-3.0 whorls. The outer surface is sculptured with numerous fine incremental striations, and the inner surface is smooth.

The radula is typically taenioglossate and consists of 45-83 transverse rows. The central tooth is 28-35μ wide. It has a basolateral extension on each side, and a tongue-like projection along the middle of the basal margin. The reflection bears a large conical mesocone and 4-6 ectocones on each side. A single basocone is located on each side of the face at the top of a reinforced crest on the basolateral extensions. The lateral tooth is clavate or hatchet-shaped, with 3 entocones, a large mesocone, and 5-7 ectocones on the reflection. The inner marginal tooth has 20-25 cusps on the reflection, and the shaft has a thin flange-like extension along the inner margin. The outer marginal tooth has about 20-30 very small slender cusps on the reflection. Its shaft has a prominent flange along the outer margin.

The foot is pandurate, rounded posteriorly, and truncate anteriorly. The snout extends slightly beyond the anterior margin of the sole, and the tentacles are about twice the length of the snout when extended. The gill consists of a series of about 20 filaments that are normal in position. The filaments are narrowest anteriorly, become progressively wider posteriorly, and taper again shortly before the pericardium. The ground color of the mantle is rich chocolate brown to bright red. It has numerous large blotches or spots that are darker than the ground color, but of the same hue.

The verge originates on the nape slightly to the right of the midline. It is dorsoventrally compressed, and when retracted it is folded back upon itself with its distal end pointing forward. When erect it extends forward and along the right side of the animal. The penis originates on the right side of the verge about halfway between the base and the apex. The penis is about equal to or greater than the length of the accessory lobe of the verge. It is spatulate and has a long narrow gland lying along each margin from near the tip to the base. The dorsal surface of the verge has a few scattered glandular tubercles over the center of the accessory lobe, or a large circular or spiral loop on the center. An apical glandular crest may be present or absent. The ventral surface has a long, narrow transverse gland extending in from the left margin toward the point of origin of the penis.

Type species: *Spilochlamys conica* new species.

Spilochlamys is known only from Florida, where it occurs through the middle St. Johns River system, and along the gulf coast from Pasco County north and west to Jackson County. It is most closely related to *Notogillia* and *Cincinnatia,* as is indicated by the structure of the verge, and the glands on the verge. It is distinguished from both of these genera by its reddish-brown spotted mantle, and by the arrangement of the glands on the verge. Its shell is distinguished from *Notogillia* by being translucent, perforate, having a non-thickened columellar margin, a sharp outer lip, and a light gray or olivaceous periostracum. Its shell is very similar to that of *Cincinnatia,* but usually may be distinguished by its more deeply impressed suture, its more strongly shouldered whorls, and its larger umbilical perforation.

Spilochlamys contains two species, which are rather remote in their relationship to each other. For this reason it is difficult to diagnose the genus, because such a diagnosis includes many major alternatives, and broad general statements. There is no doubt that the species represents a single phyletic stock. This is indicated by the reddish-brown spotted mantle, the large, flat, blade-like penis with its 2 long marginal glands, and the single narrow transverse gland on the inner surface of the verge opposite the point of origin of the penis.

S. *conica* (new species) may be identified by its narrower umbilicus, which lacks a strong circumumbilical keel, its outer lip, which lacks a callus deposit on the inner surface, and its verge, which has an apical glandular crest on the accessory lobe, and 3 short, tubercular, glandular crests on the outer surface.

S. *gravis* (new species) may be identified by its wide umbilical perforation, which is accentuated by a strong circumumbilical keel, by a thick callus on the inner surface of the lip, and by its verge, which lacks an apical glandular crest on the accessory lobe, and has a large spiral or circular glandular crest on the middle of the outer surface.

The name *Spilochlamys* is derived from the Greek σπῖλος, meaning spotted, and χλαμύς, meaning cloak. The gender of the name is feminine.

Spilochlamys conica new species

SHELL (Fig. 65, C; Fig. 57, A-C). Conical, occasionally abbreviate conical; thin or only moderately thick, transparent or translucent; light olive gray. Umbilicate, the columellar lip is indented and partially reflected over the perforation so that it tends to be rimate. A weak circumumbilical keel may be present. Spire conical or abbreviate conical, straight-sided or nearly so, usually shorter than the length of the aperture. Sculpture consisting of incremental striations which may be very distinct. Occasionally very feeble spiral striations may be evident above the periphery of the whorls. 4.1-5.4 strongly shouldered whorls (usually 4.5-5.0), regularly increasing in size and regularly descending; nuclear whorl 0.25-0.30 mm in diameter; embryonic whorls protruding, not differentiated from remaining whorls; later whorls strongly arched. Suture deeply impressed, regularly descending to the aperture. Aperture ovate or elliptical, white; upper corner rounded or weakly angulate; lower corner parabolic, usually rounded, subangulate in some specimens; columellar and parietal margins forming a deep indentation into umbilical region. Peristome complete across parietal wall; with very thin dark border along edge; outer lip rounded, usually not conspicuously thickened except at basocolumellar angle (a slight or moderate callus may occur within the outer lip in some individuals, causing the lip to appear flared); columellar margin slightly rounded or straight; parietal callus thin, shiny, opaque.

Length of shells with a complete peristome, 2.8-4.8 mm; diameter of shell, 2.4-3.4 mm; length of aperture, 1.6-2.5 mm; width of aperture, 1.3-2.0 mm; length/diameter, 1.13-1.54; length/aperture, 1.67-2.13; whorls/length, 1.00-1.38.

Measurements of type: length of shell, 4.65 mm; diameter of shell, 3.40 mm; length of aperture, 2.45 mm; width of aperture, 1.90 mm; 5.1 whorls; diameter of nuclear whorl, 0.25 mm.

The characteristics of the aperture and umbilicus are variable. Normally the outer lip is thin and lacks a callus deposit within, but occasional specimens have a slight thickening in the lip. The umbilicus usually lacks a circumumbilical keel or ridge, but some specimens have a slight or moderate rounded ridge due to thickening of the base of the body whorl in this region. In no case is it manifested as a keel that strongly accentuates the umbilicus.

Considerable differences in shell size and proportions exist between different colonies. Shells from the type locality tend to be more elongate than those from other localities, but the extremes of variation within this colony encompass those of most other colonies. Specimens from Rock Bluff Springs, Gilchrist County, are smaller, relatively more obese and thicker shelled than is common. Specimens from Blue Springs, Levy County, are similar to S. *obesa* in shell shape and umbilical width, but other features of the shell and the anatomy are normal.

Operculum (Fig. 27, D). Consisting of about two and one-quarter rapidly expanding whorls. Thin; outer surface moderately and uniformly striate. Inner surface smooth and slightly thickened over the nucleus.

Animal (Fig. 26, C). In life the sides of the foot and the anterior lateral corners of the pedicel are flecked with bright orange. The sides of the pedicel have very slight amounts of similar flecking. The eye stalk contains an intense stripe of bright orange pigmentation that extends from the tip of the tentacle to the base along the dorsal surface, and may be continuous across the nape. The snout and the lateral surfaces of the tentacles are intensely colored with melanophores. The muzzle is white. The ground color of the mantle varies from light or dark brown to intense red, and has several large spots and blotches that are darker than the ground color. White patches may occur within or between the dark spots.

Verge (Fig. 42, A, B). The penis is large, flat, blade-like, and exceeds the length of the accessory lobe. The penis bears 2 large glands. One lies along the outer margin and is visible from both the dorsal and the ventral surfaces. The gland terminates on the dorsal surface at the base of the penis. A second long narrow gland lies along the inner margin of the penis. It begins on the dorsal surface at the tip, where it may be continuous with the first gland, and twists onto the ventral surface where it usually winds beyond the base. The dorsal surface of the accessory lobe bears 4 major glands. Two lie near the middle of the left margin and are raised on low pedicels. These may be fused with each other at one or both ends. A large apical crest lies along the left margin at the tip, and is visible from both surfaces. A fourth gland lies along the right margin, slightly distal to the origin of the penis. It usually is elongate and points toward the apical crest. On the inner surface of the verge about midway between the base and the apex, and opposite the point of origin of the penis lies an elongate, transverse gland that is slightly raised on a fleshy base.

RADULA (Fig. 35, A). Five specimens examined. The ribbon contains 45-53 transverse rows. The central tooth is 28-35μ wide; its cusp formula is $\frac{5\text{-}1\text{-}5}{1\text{-}1}$ or $\frac{6\text{-}1\text{-}6}{1\text{-}1}$. The lateral tooth has 3-1-5 to 3-1-7 cusps. The inner marginal tooth has about 20 cusps. The outer marginal tooth has about 30 cusps, which are too fine to be clearly distinguished under oil immersion at a magnification of 1000 X.

TYPE LOCALITY. Natural Bridge Springs, Natural Bridge Monument, Leon County, Florida. Type: *UMMZ* 216480; collected September 6, 1963 by Fred G. Thompson. Paratypes: *UMMZ* 216481 (100).

DISTRIBUTION (Fig. 16). This species is known only from river systems draining into the Gulf of Mexico in north central Florida, from Levy County north and west to Jackson County. It may occur farther west in Florida. An isolated population from Crystal Springs, Pasco County, is tentatively referred to *S. conica.*

LOCALITIES. Florida: *Alachua Co.*—Santa Fe R., 3 mi. n.w. High Springs. *Jackson Co.*—mouth of spring run to Chipola R., e. edge of Marianna. *Gilchrist Co.*—Rock Bluff Springs, 5 mi. n.w. Bell; Bell Springs, 0.8 mi. n.w. Fanning Springs; Otter Springs, about 3 mi. w.s.w. Wilcox; Hart Springs, *Jefferson Co.*—Wacissa Springs, Wacissa. *Levy Co.*—Wekiva Springs Run, 1.1 mi. e. Gulf Hammock; Blue Springs, about 3 mi. w. Bronson; spring run 0.3 mi. s.s.e Gulf Hammock; spring run 0.2 mi. n.w. Gulf Hammock; Wacasassa R., 2.3 mi. n.e. Otter Creek; Manatee Springs, 6.0 mi. w. Hardeetown; Fanning Springs. *Pasco Co.*—Crystal Springs. *Suwannee Co.*—Charles Springs, 5 mi. n.w. Luraville; Telford Springs, Luraville. *Taylor Co.*—river 0.6 mi. n. Tennille. *Wakulla Co.*—St. Marks R., Newport; Wakulla R., at U.S. 98.

ECOLOGY. *Spilochlamys conica* was found in a variety of aquatic conditions, though usually in waters that originated from springs. Specimens are abundant in springs and spring runs, where they are usually found on or closely associated with plants over a soft sand-silt bottom. On a few occasions specimens were collected in seepage pools in hardwood swamps that were devoid of green plants. Specimens were also occasionally found in spring runs on bare sand.

FIGURE 16

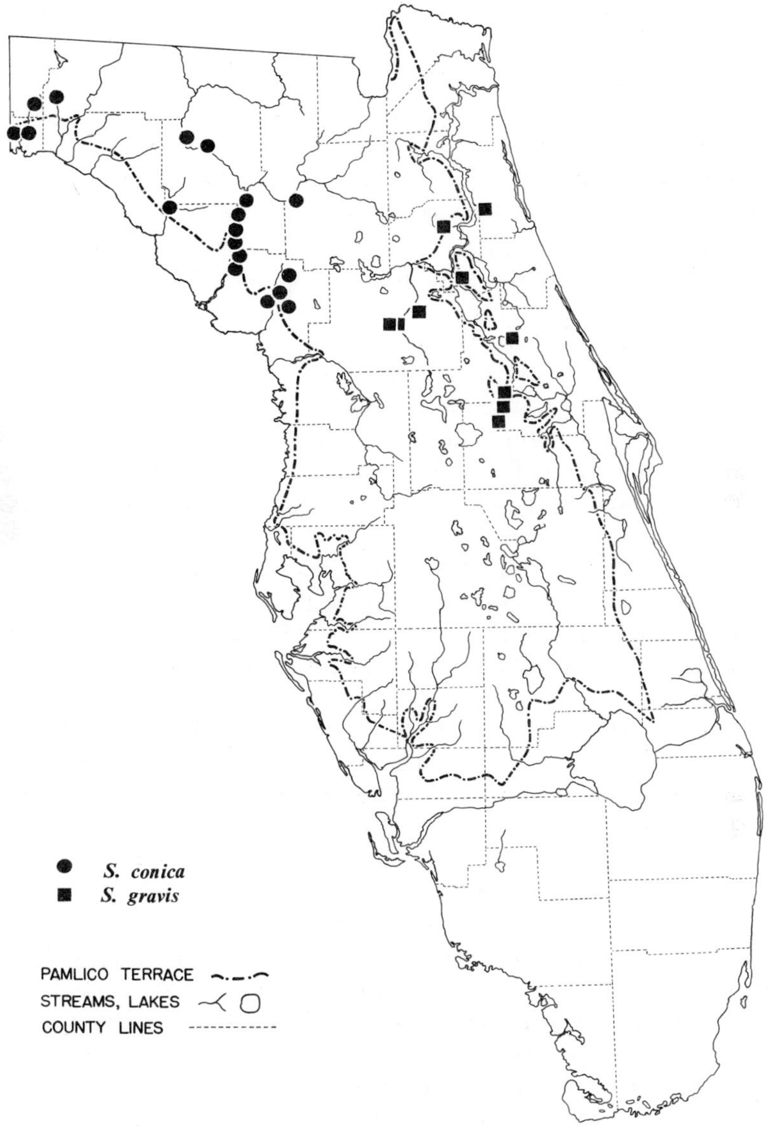

The geographic distribution of *Spilochlamys conica* new species and *Spilochlamys gravis* new species.

Spilochlamys gravis new species

SHELL (Fig. 65, A, B; Fig. 57, A-C). Ovate-globose; solid, thick, translucent or only slightly transparent (clearly transparent in live specimens). Light olive gray; very fine faint brown spiral bands may be evident, particularly near the suture of the body whorl and the aperture. These bands are about 0.05 mm wide and are about one-half the width of their intervals. Shell shiny. Openly umbilicate; perforation strongly accentuated by prominent circum-umbilical keel caused by angulation of the body whorl in that region. Spire conical, straight-sided or only slightly convex, 0.6-0.9 times the length of the aperture. Sculpture consisting of fine incremental striations. 4.4-5.0 strongly shouldered whorls; nuclear whorl 0.25-0.30 mm in diameter, protruding; remaining whorls regularly increasing in size; body whorl flattened along the periphery. Suture deeply impressed. Aperture elliptical or auriculate; white; upper corner of aperture thickened, usually forming a distinct angle inside; basocolumellar corner flaring downward, nearly angulate, shell very thick in this region; columellar and parietal margins forming a strong indentation into umbilical region. Peristome complete; outer lip rounded, sharp, with a thick white callus inside, causing lip to flare outward; columellar margin partially reflected, straight or weakly arched, continuously curved with parietal margin.

Length of shell, 3.7-5.15 mm; diameter of shell, 3.0-3.8 mm; length of aperture, 2.15-2.9 mm; width of aperture, 1.7-2.3 mm; length/diameter, 1.19-1.35; length/aperture, 1.61-1.91; whorls/length, 0.98-1.19.

Measurements of the type: length of shell, 4.65 mm; diameter of shell, 3.40 mm; length of aperture, 2.50 mm; width of aperture, 1.95 mm; 5.0 whorls; diameter of nuclear whorl, 0.25 mm.

This species is conservative in the characters that typify it. Some populations from small streams draining into the St. Johns River are composed of dwarf individuals which are otherwise typical. The width of the umbilicus may be very much reduced in size, but regardless of the size of the perforation it is always surrounded by a conspicuous keel that strongly accentuates the umbilicus.

OPERCULUM. (Fig. 27, F). Consisting of about 3 whorls. Parietal margin slightly concave or occasionally straight. Outer surface with moderately spaced delicate striations.

VERGE (Fig. 42, C, D). The penis is long, flat, and lanceolate in shape. It bears a long, narrow gland along each margin, and one or more small diffuse glands on the dorsal surface near the base. The gland along the outer margin extends from near the tip of the penis onto the base of the verge. The gland along the inner margin curves onto the under side of the penis and terminates at or near the origin of the penis. The dorsal surface of the verge has a large narrow gland that may form a complete ring or a spiral. The gland overlaps the left margin of the verge about halfway between the tip and the base, and extends diagonally to the opposite margin. Within the loop of the gland is a smaller gland that may be free from or fused to parts of the loop or spiral. The apex of the accessory lobe of the verge lacks a terminal crest. The inner surface of the verge bears a large irregular gland that may be slightly raised on a pedicel, and is opposite the point of origin of the penis. The vas deferens is simple, and lies along the right margin of the verge and the penis.

RADULA (Fig. 35, B). Six specimens examined. The ribbon contains 67-83 transverse rows. The central tooth is about 28-35μ wide. Its cusp formula is $\frac{4\text{-}1\text{-}4}{1\text{-}1}$ to $\frac{6\text{-}1\text{-}6}{1\text{-}1}$. The lateral tooth is hatchet-shaped, and has 3-1-5 or 3-1-6 cusps. The inner marginal tooth has about 20-25 cusps; the outer marginal tooth has about 20.

TYPE LOCALITY. Wekiva River at Florida State Highway 46, 9 miles west of Sanford, Seminole County, Florida. Type: *UMMZ* 216482; collected March 3, 1964 by Fred G. Thompson. Paratypes: *UMMZ* 216483 (61), 216484 (45), *MCZ* 258006 (15), *USNM* 636945 (18), *ANSP* 301542 (12).

DISTRIBUTION (Fig. 16). This species is confined to north central Florida, where it has been collected in the St. Johns drainage system from Palatka south to the Wekiva River.

LOCALITIES. Florida: *Marion Co.*—Silver Springs; Oklawaha R., 8.5 mi. e.n.e. Ocala; creek on Fla. 314, 6 mi. n.e. Grahamsville. *Putnam Co.*—Welaka Springs, 1 mi. n. Welaka; Rice Creek, 7.3 mi. w.n.w. Palatka. *Seminole Co.*—Wekiva R., 2 mi. s. Fla. 46; Wekiva R., 3 mi. s. Fla. 46; Wekiva R. at Fla. 46. *St. Johns Co.*—creek 3.1 mi. n.n.w. Spuds. *Volusia Co.*—Lake Woodruff at De Leon Springs.

ECOLOGY. *Spilochlamys gravis* is found in soft or silty sand in streams and rivers that usually originate from springs. Vegetation is usually present in these streams, but the snails are not restricted to the vicinity of the plants.

Cincinnatia Pilsbry, 1891

Cincinnatia Pilsbry, 1891, Proc. Acad. Nat. Sci. Phila.: 327.—Baker, 1928, Bull. Wisc. Geol. Nat. Hist. Surv., 70 (1): 123.—Berry, 1943, Misc. Publ. Museum Zool. Univ. Mich., (57): 32.

Type species by original designation: *Paludina cincinnatiensis* Anthony, 1841 (=*Paludina integra* Say, 1821).

Shell simple, conical or cylindric-conical; of moderate size; umbilicate or imperforate. Spire conical, 0.8-1.7 times the length of the aperture, straight-sided or convex. Sculpture consisting of fine or moderate incremental striations; occasionally spiral striations on thin shells may be detected with transmitted light. 3.9-5.7 whorls in mature shells; nuclear whorls nearly depressed or moderately elevated, 0.20-0.50 mm in diameter; remaining whorls varying from nearly flat-sided to strongly arched, occasionally shouldered. Aperture ovate to elliptical; upper corner angulate or rounded; lower corner angulate or rounded at basocolumellar margin; axis of aperture at 30-45° to axis of shell; plane of aperture parallel to or at a slight angle to axis of shell. Peristome complete at maturity; outer lip rounded, sharp, straight-edged or nearly so in lateral view; columellar margin weakly reflected, partially obscuring or closing umbilicus, thin, usually sharp-edged; parietal margin thin or moderately thick. Inside of aperture very seldom with a thick callus deposit.

Length of shell, 2.00-4.9 mm; length/diameter, 1.23-1.84; length/aperture, 1.75-2.73; whorls/length, 1.00-2.00.

The operculum is ovate, membranous, or rigid; colorless or bright amber, transparent. Paucispiral to nearly multispiral, consisting of 2 to 3 whorls. The nucleus is subcentrally located in the lower and left third of the operculum. The outer surface is shiny, and has few to many fine incremental striations. The inner surface is smooth.

Anatomical data are based on the type species (Berry, 1943: 33-36) and the Florida species. The foot is short, spatulate, bluntly rounded posteriorly and truncate anteriorly. The anterior margin of the sole is richly supplied with mucous glands. The snout extends only slightly beyond the anterior margin of the sole. The tentacles are short and tapered. When extended they reach only slightly beyond the tip of the snout. Pigmentation may be very intense or nearly lacking. The species usually have a dark band around the muzzle, and some dark shading on the sides of the pedicel and the foot. The mantle may be spotted or shaded with melanophores, or it may be unmarked. The gills consist of a long series of about 30-40 filaments, which are of nearly uniform width. The filaments near the mantle collar and the pericardium are tapered.

The verge (Fig. 26, D) originates on the nape at a distance behind the tentacles equal to the length of the snout. When retracted the verge is folded upon itself with its distal end pointing forward and to the right. When erect the verge extends from beneath the mantle and along the right side of the animal. The verge is broad and dorsoventrally flattened. The penis originates on the distal half along the right margin and is a small, slender, conical structure. A large number of glandular tubercles and crests are scattered over the surfaces of the verge and the penis, and a series of terms is presented to facilitate their discussion (Fig. 17). The secondary lobe of the verge terminates in a heavy glandular crest called the *apical crest.* A bulb-like protuberance, the *terminal bulb,* extends dorsolaterally from the apical crest toward the penis, and a small patch of glands is usually present at the tip of the bulb. Scattered over the midregion of the outer surface of the verge and the penis are a number of small glandular tubercles and crests, collectively called the *superior tubercles.* They may occur as a few small scattered tubercles, or they may be fused into a series of long slender crests. On the inner surface of the verge and slightly behind the apical bulb is a transverse glandular crest called the *parapical crest.* This crest may be very heavy and raised on a fleshy pedicel. Slightly distal to the parapical crest is a smaller glandular crest, the *accessory crest.* Lying across the verge about halfway between the base and the parapical crest is another glandular crest that is raised on a fleshy base, the *inferior crest.* The accessory crest and the inferior crest are absent in some species. All species possess in some modified form the other glandular processes.

The radula is typically taenioglossate. The teeth are arranged in 29-70 transverse rows. The central tooth is 32-48μ wide. Its basolateral corners extend outward as two peg-like projections that have a reinforced crest along the face. The basal margin has a tongue-like median projection. The reflection has a long slender mesocone and 3-6 ectocones on each side. A single basocone is present on each side of the face of the central tooth at the top of the crest on the basolateral extension. The lateral tooth is typically clavate. It has a large rectangular body and a long lateral shaft. The reflection bears 2-4 entocones, a large mesocone, and 3-9 ectocones. The inner marginal tooth has 13-20 cusps along the reflection. The base of the shaft has a moderate flange-like projection along its inner margin. The outer marginal tooth has about 20-25 very small slender cusps on the reflection. The cusps are too small to be clearly differentiated with magnifications of 1000 X. The shaft is about twice the length of the distal end, and has a long, wide flange along its outer margin.

FIGURE 17

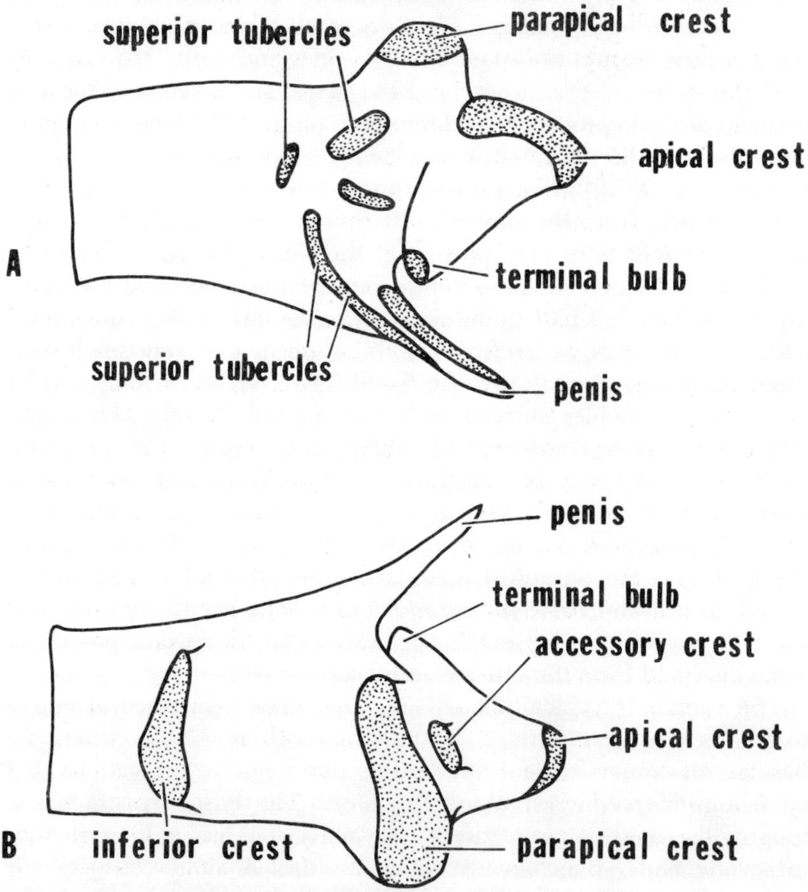

A generalized verge of *Cincinnatia* showing the glandular tubercles and crests. A. The outer surface. B. The inner surface.

The radulae of all of the Florida species of *Cincinnatia* are monotonously alike. They have no distinctive features at the specific level.

Cincinnatia is allied to *Somatogyrus, Birgella, Notogillia,* and *Spilochlamys* by the general structure of its verge. It differs conchologically from all these genera by having a simple conical shell with a simple, complete peristome, and by lacking any modification of the shell in the region of the umbilicus or the aperture. Its verge differs from that of the others by having the glands arranged as described above.

Cincinnatia is confined to the North American continent, where it is known from the Great Lakes region south to Florida and Texas. It is not known how far west the genus extends, but some of the species described as *Amnicola* from west of the Mississippi River probably belong here. The only species that have definitely been demonstrated to belong in *Cincinnatia* are the type species and those dealt with in this paper.

All previous authors who have discussed *Cincinnatia* have failed to note its most distinctive feature, which is the presence of glandular processes on the verge. Pilsbry (1891, 327) defined *Cincinnatia* in a very brief footnote as having a smaller radula than does *Amnicola.* F. C. Baker (1928, 123) separated *Cincinnatia* from *Amnicola* because of its protruding nuclear whorls as opposed to the depressed nuclear whorls of *Amnicola.* He erroneously stated that the verge was like that of *Amnicola.* Thiele (1931, 139-40) used the characters emphasized by Pilsbry and Baker. Berry (1943, 32) followed the arrangement established by Pilsbry and Baker, and pointed out the distinctive shape of the verge, but did not discuss the development of the glands.

Cincinnatia contains two species groups in Florida. The groups are distinguished by the development and distribution of the glands on the verge. The *floridana* group is characterized by a few simple superior tubercles on the outer surface, by the parapical crest raised on a low ridge, and by the lack of an accessory crest and an inferior crest on the inner surface. The *vanhyningi* group is characterized by having superior tubercles that are fused and extended into crests that form a circular or U-shaped pattern on the outer surface, by the parapical crest raised on a fleshy pedicel, and by an accessory crest and an inferior crest on the inner surface of the verge.

The *floridana* Group

This group of four species is characterized by the presence of only a few superior tubercles on the outer surface of the verge, and by lacking an accessory crest and an inferior crest on the inner surface of the verge. The species are distinguished as follows.

C. floridana (Frauenfeld). Shell medium-sized, 2.8-3.5 mm long; moderately thick and opaque or only slightly translucent. Spire conical, 0.9-1.1 times length of aperture. 4.2-4.7 whorls. Penis with glandular crests.

C. monroensis (Dall). Shell moderately large, 3.5-4.7 mm long; thick and opaque. Spire conical, 1.00-1.35 times length of aperture. 4.8-5.4 whorls. Penis with glandular crests.

C. mica (new species). A diminutive version of *C. floridana.* Shell 2.0-2.3 mm long; thin and transparent. Spire conical, 0.7-1.1 times length of aperture. 3.9-4.2 whorls. Penis lacking glandular crests.

C. helicogyra (new species). Shell relatively large and cylindric-conical; 4.0-4.9 mm long; transparent. Spire 1.3-1.7 times length of aperture, convex in outline. 5.0-5.7 strongly rounded, shouldered whorls. Penis like that of *C. floridana,* but with fewer tubercles.

Cincinnatia floridana is generally distributed throughout the northern half of the peninsula as far north and west as the Suwannee River. *Cincinnatia mica* is confined to a single spring in the Ichetucknee River adjacent to typical populations of *C. floridana.* *Cincinnatia helicogyra* is confined to a single spring area at the head of Crystal River. *Cincinnatia monroensis* is confined to a single seepage run entering Lake Monroe, at Enterprise, Florida.

Cincinnatia floridana (Frauenfeld)

Amnicola floridana Frauenfeld, 1863, Verhandel Kais. Konig. Zool. Botan. Ges. Wein, 13: 1028.–*ibid.*, 1865: 529; pl. X.–Dall, 1885, Proc. U.S. Nat. Museum, 8: 257-58.

Amnicola floridana var. *convexa* Pilsbry, 1892, *in* Dall, Trans. Wag. Free Inst. Sci., 3(2): 338. Type locality: Pliocene marl of the Caloosahatchee Formation and Shell Creek, Florida. (Not *Amnicola convexa* Sandberger, 1874).

Amnicola sanctijohannis Pilsbry, 1899, Naut., 13: 20.–*ibid.*, 64: 120; pl. 9, figs. 2, 2a. Type locality: St. Johns River, Astor, Lake County, Florida.

Amnicola augustina Pilsbry, 1904, Naut., 17: 113-14. Type locality: a small creek near St. Augustine, St. Johns County, Florida.

Amnicola floridana convexior Wenz, 1924, Senckenbergiana, 6: 221-22. (New name for *A. f. convexa* Pilsbry).

SHELL (Fig. 66, C, D). Conical; translucent or opaque; whitish or of varying shades of brown. Umbilicus open, rimate or closed. Spire conical, convex in outline, 0.90-1.15 times the length of the aperture, usually narrow. Surface shiny in clean shells, nearly smooth; sculpture consisting of discontinuous, fine incremental striations. 4.2-4.7 whorls in shells with complete peristomes; embryonic whorls protruding; nuclear whorl flat-topped, 0.25-0.50 mm in diameter; remaining whorls regularly increasing in size. Suture moderately or deeply impressed and regularly descending to the last whorl, where it descends slightly more rapidly. Aperture ovate; interior hyaline or white; upper corner angulate or bluntly rounded; in specimens from lotic habitats it may be broadly rounded, and the general shape may be more ample; axis of aperture lying at about 30-45° to axis of shell; plane of aperture lying at about 5-10° to shell axis. Peristome complete; usually adnate to preceding whorl; outer lip and basal lip evenly rounded with columellar margin; outer lip nearly straight in lateral view, extending very slightly forward near middle, and slightly indented between upper corner and middle.

The following measurements and ratios apply to about 90 per cent of the specimens that I have examined. Occasional dwarf populations with the same relative proportions and number of whorls may be found, particularly in springs and seepage ditches. Length of shell, 2.8-3.5 mm; diameter of shell, 2.0-2.3 mm; length of aperture, 1.4-1.7 mm; width of aperture, 1.1-1.4 mm; length/diameter, 1.35-1.50; length/aperture, 1.90-2.15; whorls/length, 1.35-1.55.

Measurements of lectotype: length of shell, 2.70 mm; diameter of shell, 1.98 mm; length of aperture, 1.37 mm; width of aperture, 1.20 mm; 4.2 whorls; diameter of nuclear whorl, 0.27 mm; peristome free from preceding whorl.

Cincinnatia floridana inhabits a large variety of habitats, and shows considerable ecological variation. Specimens from sloughs, ponds, ditches, and lakes in which there is little or no current or wave action have a thin, brown, obese shell, a rounded, ample aperture that frequently is free from the preceding whorl, and a deeply impressed suture. Specimens from springs, streams, and lakes in which there is a current or wave action have a thicker, lighter colored, more slender shell, an angulate, narrower aperture that is closely appressed to the preceding whorl, and a less impressed suture. The perforation of the umbilicus is dependent upon the relationship of the aperture to the preceding whorl. Populations representing either extreme of variation, or all intermediate stages, are common. Most population samples from the St. Johns, Withlacoochee, and Suwannee River systems contain specimens representing either extreme, as well as intermediate forms.

A peculiar form occurs at Weeki Wachee Springs, Hernando County, Florida, in which the apical whorls may become very loosely attached or free from each other.

Cincinnatia floridana, as was described by Frauenfeld, and as is represented by the cotypes, is the form commonly found in lotic habitats. Pilsbry (*in* Dall, 1892: 338) named the variety *convexa* (=*convexior* Wenz) on the basis of characters of the shell, which, he stressed, were too variable to be given much emphasis. Pilsbry (1899, 20) distinguished *sanctijohannis* by the characters that are typical for specimens from lentic habitats. In light of the material that I have examined, *convexior* and *sanctijohannis* can be considered as only ecological variations of *floridana*, and of no nomenclatorial significance.

OPERCULUM (Fig. 31, B). Ovate, membranous, chitinous, amber-colored, transparent; upper corner bluntly pointed; outer and basal margin evenly rounded; columellar and parietal margin nearly straight, and forming a rounded angle at their intersection. Consisting of 3-3.5 whorls; nucleus subcentral, located nearly halfway between the columellar margin and the outer lip margin, and at about one-third of the distance from the base to the apex. Outer surface dull, with distinct incremental striations; inner surface smooth and shiny.

ANIMAL (Fig. 26, D). The foot is spatulate. The anterior margin is bluntly indented and richly supplied with mucous glands. The sides are constricted near the anterior border of the pedicel, and extend outward posterior to the pedicel. The snout extends only slightly beyond the anterior margin of the foot, and the extended tentacles are only slightly longer than the snout. Pigmentation is usually confined to a few scattered melanophores on the sides of the pedicel near its anterior margin, and a diffuse band around the snout. This band may be interrupted middorsally. The penis is very dark in some specimens. The rest of the verge is not colored.

VERGE (Fig. 43, A-E). The verge is 1.5-2.3 mm long. The penis is 0.3-0.5 mm long. Usually there are 2 or 3 superior tubercles behind the apical bulb near the left margin. A single superior tubercle is usually located on the middle of the verge near the right margin. Frequently an additional glandular crest occurs on or near the base of the penis. Usually the superior tubercles are oval or bean-shaped. Occasionally they are elongate, and some tubercles along the right margin may be fused or extended into long narrow crests.

RADULA (Fig. 33, B). Ten specimens examined. The teeth occur in 40-48 rows. The central tooth is 32-42μ wide. The cusp formula is $\frac{4\text{-}1\text{-}4}{1\text{-}1}$ to $\frac{5\text{-}1\text{-}5}{1\text{-}1}$. The lateral tooth has 2-1-4 cusps. The inner marginal tooth has 15-18 cusps.

TYPE LOCALITY. East Florida. Lectotype of *Amnicola floridana*: Vienna Naturhistorisches Museum; selected from a vial containing six other specimens; two other vials, containing two and five specimens respectively, also belong to the type series in Vienna. Lectotype of *Amnicola sanctijohannis*: *ANSP* 69461; besides containing the species with which Pilsbry intended the name to be associated, the cotypes also include specimens of *Littoridinops monroensis* and *Hyalopyrgus aequicostatus* (males). Lectotype of *Amnicola augustina*: *ANSP* 58088. I have not seen topotypic anatomical material pertaining to this nominate species; thus its status is still open to question.

DISTRIBUTION (Fig. 18). This species apparently is confined to Florida. It is distributed over the northern half of the peninsula as far north and west as the Suwannee River. Through the central part of the state it is found as far south as the area around Orlando, while on the west coast it has been found in springs and streams as far south as Hillsborough County.

Goodrich's records of *floridana* and *sanctijohannis* (1944, 6) from Alabama are erroneous. His lots of both forms include a species of *Cincinnatia* and a species of *Amnicola*, both of which are undescribed. Dall (1885, 257) recorded *floridana* from Key West. Apparently this record was based upon a misidentification of *Littoridinops monroensis*.

LOCALITIES. Florida: *Alachua Co.*—spring 2.3 mi. w. High Springs; Hornsby Springs, 3 mi. n. High Springs. *Citrus Co.*—Withlacoochee R. below dam of reservoir; Withlacoochee R., 4 mi.

FIGURE 18

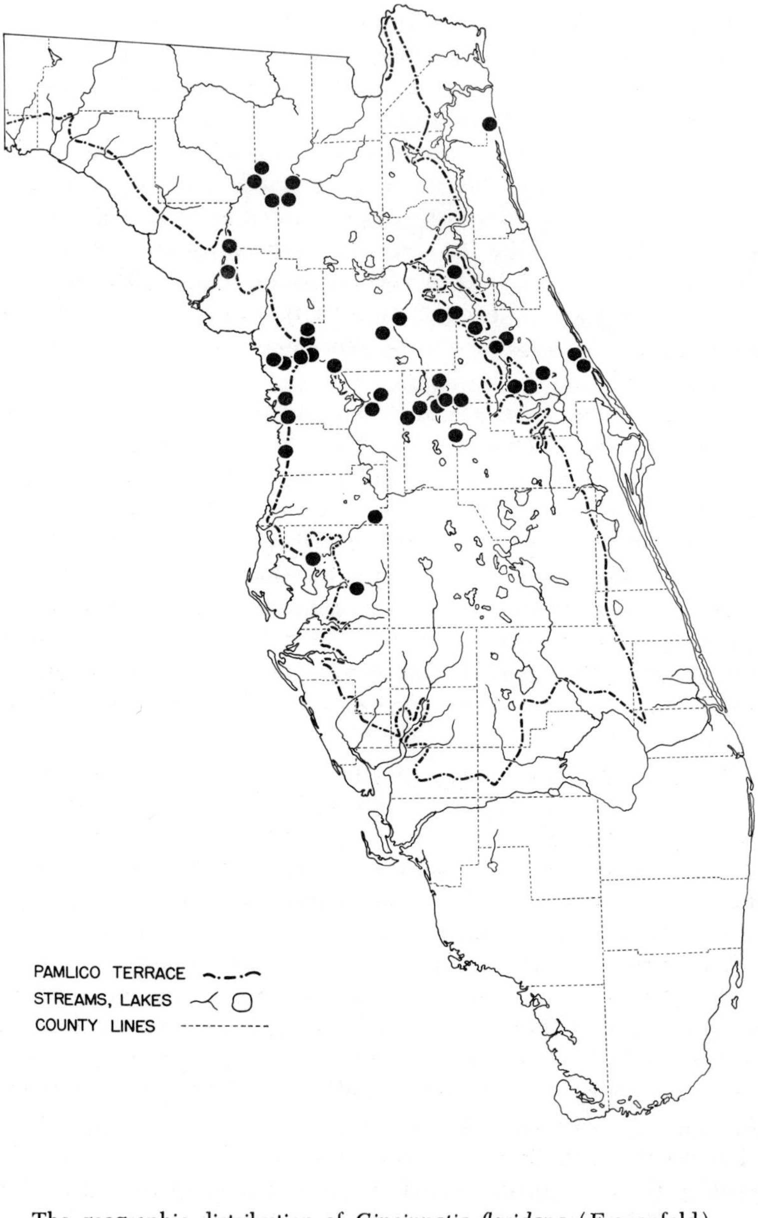

The geographic distribution of *Cincinnatia floridana* (Frauenfeld).

above dam; Withlacoochee R., at Fla. 200; Withlacoochee R., Dunnellon; Homosassa Springs; small spring 0.5 mi. s.w. Homosassa Springs; Chassahowitzka R., Chassahowitzka. *Columbia Co.*—Ichetucknee Springs. *Gilchrist Co.*—Blue Springs, 4.5 mi. w. High Springs. *Hernando Co.*—Weeki Wachee Springs. *Hillsborough Co.*—Sulphur Springs, Tampa; Lithia Springs, 3 mi. w.s.w. Lithia. *Lake Co.*—Bugg Springs, 0.3 mi. n.w. Okahumpka; Blue Springs, 1 mi. w. Yalaha; Lake Harris at Fla. 19; Lake Apopka, Montverde; Lake Dora, 1 mi. s. Mt. Dora; Lake Eustis, Grand Island; Lake Eustis, Eustis; St. Johns R., Astor. *Levy Co.*—Withlacoochee R., Ingles; Fanning Springs; Manatee Springs, 6.0 mi. w. Hardeetown. *Marion Co.*—Juniper Springs; Juniper Springs Run at Fla. 19; Oklawaha R., 8.5 mi. e.n.e. Ocala; Silver Springs; Rainbow Springs; Rainbow Spring Run, 2.8 mi. n. Dunnellon; Rainbow Spring Run, 1 mi. e. Dunnellon. *Pasco Co.*—Crystal Springs. *Putnam Co.*—Welaka Springs, 1 mi. n. Welaka; Mud Springs, 1 mi. s. Welaka. *St. Johns Co.*—river 0.2 mi. s. Spuds; lake on Fla. 210-A, Ponte Vedra Beach. *Sumter Co.*—Fenney Springs, 2 mi. e. Coleman; creek 2 mi. s. Coleman. *Volusia Co.*—Smyrna (*UMMZ* 69824); creek nr. Smyrna (*UMMZ* 119912); Ponce de Leon Springs; Lake Woodruff; Bay Creek, 5 mi. e. Osteen; Padgett's Spring, 1 mi. e. DeBary; Green Springs, 1 mi. s.e. Enterprise.

ECOLOGY. *Cincinnatia floridana* is found in springs, streams, ditches, sloughs, seepage pools, and lakes. The largest populations encountered were in springs and spring runs, but they are also abundant in other habitats, especially where vegetation is favorable.

Specimens are usually found on aquatic plants, preferably rooted angiosperms and floating forms such as hyacinths and water lettuce. Occasional specimens may be found foraging on the bottom, particularly where algae and diatoms have accumulated. The substrates from which snails have been collected include sand, gravel, rock, and silt. On all occasions the water was clear enough to permit visibility of the bottom.

Cincinnatia monroensis (Dall)

Bythinella monroensis Dall, 1885; Proc. U.S. Natl. Museum, 8(17): 256-57; pl. 17, fig. 9.

Paludestrina monroensis (Dall), Walker, 1918; Misc. Publ. Museum Zool. Univ. Mich., (6): 138.

SHELL (Fig. 66, A, B). Medium-sized, conical, 1.38-1.60 times as long as wide; moderately thick, translucent or opaque. Spire 1.00-1.35 times the height of the aperture, straight-sided. Umbilicus narrowly perforate, about one-sixth to one-eighth the diameter of the shell, partially shielded by the peristome. Periostracum light olive brown, usually covered with a thin rusty mineral deposit. Surface smooth or sculptured with a few fine incremental striations that may become more distinct near the umbilicus. 4.8-5.4 whorls, 1.15-1.36 whorls/length; nuclear whorl 0.20-0.30 mm in diameter, planular, protruding; remaining whorls moderately arched, not shouldered; suture moderately or deeply impressed. Aperture ovate or rhomboid; upper corner of aperture angulate; axis of aperture lying at about 30-40° to axis of shell; plane of aperture lying at about 10° to shell axis; interior of aperture livid white but may appear reddish due to transmission of light from rust on periostracum. Peristome complete across parietal margin; in contact with preceding whorl; outer lip and basal lip evenly rounded, sharp-edged with a slight internal callus at maturity; basocolumellar corner occasionally projected forward resulting in a rhomboid shape of the aperture; columellar margin indented into umbilical region, dull-edged; parietal margin nearly straight, with a moderately thick callus.

Length of shell, 3.5-4.7 mm; diameter of shell, 2.4-3.4 mm; height of aperture, 1.6-2.1 mm; width of aperture, 1.4-1.9 mm.

Measurements of lectotype: length of shell, 4.50 mm; diameter of shell, 3.00 mm; height of aperture, 2.00 mm; width of aperture, 1.60 mm; 5.3 whorls; nuclear whorl 0.20 mm in diameter.

The specimens in the type series are larger in average size than are specimens collected during recent years. There is a broad overlap, and some recent specimens are as large as specimens of the type series. The type lot differs further in that the aperture is more frequently rhomboid than it is in current specimens. These slight differences may be due to genetic drift of the population, or they may be due to ecological changes that have been brought about through modification of the habitat.

OPERCULUM (Fig. 27, A). Ovate, apex bluntly pointed; chitinous, transparent, amber in color. Paucispiral, consisting of about 3 rapidly expanding whorls. Nucleus located at about one-fourth of the distance from the base of the apex and at about two-fifths of the distance from the left margin to the right. Outer surface sculptured with a few fine incremental striations. Inner surface smooth, with a small knob over the nucleus.

VERGE (Fig. 44, A, B). The verge is similar to that of *C. floridana*. The apical crest and the parapical crest are considerably enlarged. The parapical crest nearly extends across the verge and is moderately raised on a fleshy pedicel. One or 2 small accessory crests may be present distal to the inner end of the parapical crest. There are 1 or 2 short, stout, superior tubercles located above and behind the terminal bulb. Occasionally these tubercles may be slightly raised on thin fleshy pedicels. Another small, round, superior tubercle occurs near the right margin about halfway between the base of the verge and the base of the penis.

RADULA. The teeth occur in 62-64 transverse rows. The central tooth has 4-5 entocones and a single basocone on each side of the mesocone. The lateral tooth has 2 entocones, a single large mesocone, and 3-4 ectocones. The inner marginal tooth has 17-19 cusps and the outer marginal has about 20 cusps.

TYPE LOCALITY. The type locality was the brook flowing from Benson's Mineral Spring, at Enterprise, Volusia County, Florida. Recently the spring has been capped as a water source for a hydroelectric plant, and all that remains of the original habitat is a small seepage run that originates from a few small seepage springs in a low wooded lot nearby. This will soon be eliminated with modern "improvements" taking place in the area. Lectotype: *USNM* 37502. Paratypes: *USNM* 636130 (30); *UF* (30) ex. C. T. Simpson collection. The lectotype is the specimen figured by Dall.

DISTRIBUTION (Fig. 19). Confined to the type locality.

ECOLOGY. Live specimens are fairly common in soft, silty ooze and on aquatic debris in the seepage run. No submerged vegetation other than diatoms grow in the run because of the dense shade produced by the thick, damp woods.

Cincinnatia mica new species

SHELL (Fig. 66, E). Obese conical; thin, transparent; light gray. Umbilicus rimate, columellar lip reflected over but not occluding perforation. Spire 0.75-1.05 times the length of the aperture; sides of spire convex, occasionally straight-sided. Sculpture consisting of fine but distinct incremental striations. 3.9-4.2 whorls; nuclear

FIGURE 19

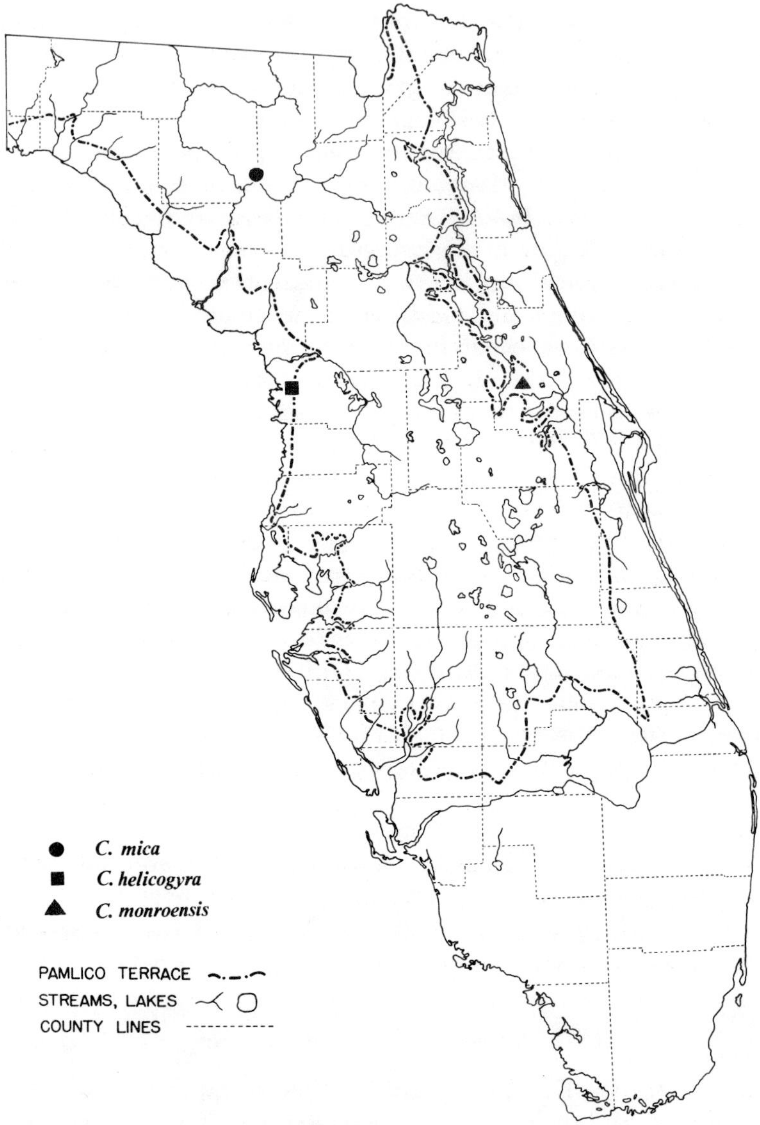

The geographic distribution of *Cincinnatia mica* new species, *Cincinnatia helicogyra* new species, and *Cincinnatia monroensis* (Dall).

whorl 0.25-0.30 mm in diameter, depressed, only slightly elevated above following whorl; remaining whorls of spire inflated; body whorl slightly arched above, evenly rounded at periphery and below. Suture deeply impressed. Aperture ovate; interior hyaline; upper corner usually angulate at about 90°; axis of aperture lying at about 45° to axis of shell; plane of aperture lying about 10-20° to shell axis. Peristome complete in large specimens, although it is very thin and tightly adnate over the parietal wall; outer lip thin, fragile, straight-sided or very slightly arched forward in lateral profile; outer lip may extend very slightly forward at insertion; outer lip, basal lip, columellar margin, and parietal margin continuously rounded; shell noticeably thickened at basocolumellar corner; parietal margin following curvature of preceding whorl, arched slightly beyond plane of aperture.

Length of shell, 2.00-2.30 mm; diameter of shell, 1.50-1.80 mm; length of aperture, 1.10-1.20 mm; width of aperture, 0.90-1.00 mm; length/diameter, 1.23-1.43; length/aperture, 1.75-2.05; whorls/length, 1.80-2.00.

OPERCULUM (Fig. 31, E). Ovate, membranous, colorless. Nearly multispiral, consisting of about three and one-half slowly expanding whorls; nucleus subcentrally located about halfway between columellar margin and outer lip margin, and at about one-third of the distance from the base to the apex. Outer surface with a few faint incremental striations; inner surface smooth.

VERGE (Fig. 43, F-I). The penis is relatively slender and extends nearly to the tip of the apical crest. Usually there are 1-3 small superior tubercles, but a penial crest is lacking. The apical crest is normal. The parapical crest is raised on a low ridge, and usually extends across the verge. An accessory tubercle and an inferior pedicel are lacking.

RADULA. Five specimens examined. The teeth are arranged in 29-49 transverse rows. The central tooth is 36-42μ wide; its cusp formula is $\frac{4\text{-}1\text{-}4}{1\text{-}1}$ or $\frac{5\text{-}1\text{-}5}{1\text{-}1}$. The lateral tooth has 2-1-3 cusps. The inner marginal tooth has 13-16 cusps.

TYPE LOCALITY. A small spring along the west bank of the Ichetucknee River about 1 mile northeast of U.S. Highway 27, Suwannee County, Florida. Type: *UMMZ* 216494; collected March 7, 1963 by Fred G. Thompson. Paratypes: *UMMZ* 216495 (200), *USNM* 636939 (30), *ANSP* 301538 (30), *MCZ* 258012 (35), *CM* 65-9 (50).

DISTRIBUTION (Fig. 19). Known only from the type locality.

ECOLOGY. The spring at the type locality issues from two sources from beneath a sandstone outcropping. The spring forms a large circular pool that is about 40 feet wide and 1 or 2 feet deep. The pool is open and continuous with the Ichetucknee River. The bottom of the pool is covered with sand and gravel, and there are moderate deposits of silt along the shore. Vegetation includes large patches of submerged bryophytes that cover sticks and rocks, and large clumps of cypress rootlets.

The Ichetucknee River at this point is 3-5 feet deep, and has a current of about 10 miles per hour. The bottom of the river is covered with gravel and has large patches of eel grass (*Vallisneria* sp.).

Cincinnatia mica was found only on bryophytes and cypress rootlets in the pool, where it was associated with *Goniobasis floridensis* and *Notogillia wetherbyi*. *Cincinnatia floridana* was very abundant in the river immediately adjacent to the spring pool, but did not occur in the pool.

***Cincinnatia helicogyra* new species**

SHELL (Fig. 68, D). Elongate conical, approaching cylindric-conical; moderately thick; barely transparent or else translucent; light olivaceous gray. Umbilicate; columellar margin of aperture partially reflected over umbilicus, but not obscuring it. Spire attenuate, about 1.3-1.7 times the length of the aperture; sides of spire straight or weakly convex. Sculpture consisting of numerous close, fine, but distinct, incremental striations that become more intense near the umbilicus; these in turn are crossed by numerous very fine, weakly incised spiral striations, which are most distinct near the shoulder of the body whorl. 5.0-5.7 whorls; nuclear whorl protruding, 0.20-0.30 mm in diameter; remaining whorls inflated, strongly shouldered. Suture deeply impressed. Aperture basally located; ovate or elliptical; interior of aperture and parietal callus white; upper corner of aperture rounded in specimens with complete peristome; axis of aperture lying at about 20-30° to axis of shell; plane of aperture lying at about 10-20° to shell axis. Peristome complete across parietal margin only in large specimens; basocolumellar area thickened; outer lip thin and fragile straight-edged in lateral view; columellar margin weakly reflected, thin; parietal callus straight and lying within plane of aperture.

Length of shell, 4.0-4.9 mm; diameter of shell, 2.4-2.8 mm; length of aperture, 1.6-1.9 mm; width of aperture, 1.3-1.6 mm; length/diameter, 1.65-1.84; length/aperture, 2.30-2.73; whorls/length, 1.15-1.35.

Measurements of type: length of shell, 4.7 mm; diameter of shell, 2.7 mm; length of aperture, 1.9 mm; width of aperture, 1.5 mm; 5.5 whorls; diameter of nuclear whorl, 0.25 mm.

OPERCULUM (Fig. 31, C). Ovate, thin, chitinous, transparent, amber colored. Paucispiral, consisting of two and one-half rapidly expanding whorls; nucleus located in lower and left third of operculum. Outer surface dull and intensely sculptured with fine incremental striations. Inner surface smooth, shiny, noticeably thickened over early nuclear whorls.

VERGE (Fig. 45, A-D). The development of the penis, terminal bulb, apical and parapical crests are as in *C. floridana*, but the outer surface usually bears only 1 or 2 superior tubercles. The maximum development of the verge occurs in specimens with a shell 3.0-3.7 mm long. Larger specimens have a noticeably reduced verge, and in very large males it may be a simple knobby flagellum.

TYPE LOCALITY. A large spring-fed lagoon on the south side of the head of the Crystal River, Citrus County, Florida. Type: *UMMZ* 216496; collected August 24, 1962 by Fred G. Thompson. Paratypes: *UMMZ* 216497, 216498.

DISTRIBUTION (Fig. 19). Known only from the type locality.

ECOLOGY. The type locality is a large lagoon about one-fourth of a mile long and 200 yards wide. It drains from the south into the Crystal River by way of a channel about 50 feet wide. There are several large springs along the periphery of the lagoon and discharging below the surface of the water. Although the lagoon and the Crystal River are spring fed, the water may be slightly brackish at times due to tidal fluctuations, which may be as much as 2 feet.

The bottom of the lagoon is covered with fine sand that contains large amounts of decayed organic material. The whole lagoon is covered by a thick bed of eel grass (*Vallisneria* sp.) and small interspersed patches of other vegetation, particularly *Elodea*. Snails were scarce on the three occasions that I visited the spring. They were found primarily on *Elodea*, and to a lesser extent on eel grass, and only near a large boil at the head of the channel where they were associated with *Littoridinops monroensis* and *Hyalopyrgus aequicostatus*.

The *vanhyningi* Group

This group consists of six species that have typical *Cincinnatia* shells with a conical spire. The distinguishing feature of the group involves the development of the glands on the verge. In some species the superior tubercles are fused and extended into long, narrow crests that are arranged in a rather consistent pattern. A central series of crests coalesce to form a prominent U-shaped gland on the middle of the verge. The ends of the U may come together to form a complete ring, and/or the U may be broken into segments. In all cases the basic pattern of the U is present. Additional crests and tubercles may lie within and parallel to the arms of the U. These crests and tubercles may coalesce with the ends of the U to form an irregular spiral, and they may occasionally extend completely to the base of the U and form a flattened E-shaped crest. A series of prominent crests lies along the periphery of the U, and may be continuous with the glandular crests along the outer margin of the penis. In other species the superior tubercles may be arranged in circular or longitudinal series but may not form a distinct U. The parapical crest is well developed, and is raised on a thick, fleshy pedicel. An accessory crest may be present on the inner surface of the verge distal to the parapical crest. The accessory crest is usually raised on a narrow pedicel, but may be reduced to a small tubercle or it may be absent. A very prominent inferior crest is present in all but one species. In that species it is reduced to a small tubercle or may be lacking.

This group of species is confined to the lower St. Johns River system and one of its tributaries, the Wekiva River. Five of the species are confined to single springs. The sixth species is confined to small streams along the lower St. Johns River. Comparisons of the six species are given in Table 5. They may be distinguished by the following characters.

C. vanhyningi (Vanatta). Shell small, thick, opaque, with a shallow suture. Lower corner of aperture tending to be angulate and to have a fluted channel. Operculum nearly multispiral with 3 large, slowly expanding whorls. The verge lacks an accessory crest, and there are no crests within the U-shaped superior tubercle.

C. ponderosa (new species). Shell large, relatively few whorls per length; suture deeply impressed, causing whorls to be globose. Operculum paucispiral, with 2 rapidly expanding whorls. Verge

TABLE 5.—SHELL MEASUREMENTS OF THE *Cincinnatia vanhyningi* GROUP OF SPECIES, BASED ON THIRTY-FIVE SPECIMENS OF EACH SPECIES. MEASUREMENTS IN MILLIMETERS

Species	Length	Diameter	Aperture height	Aperture width	Whorls	Whorls / Height
ponderosa	4.0 - 4.6	2.8 - 3.1	2.1 - 2.3	1.7 - 1.85	4.3 - 4.9	1.0 - 1.15
wekiwae	3.1 - 3.7	2.1 - 2.5	1.6 - 1.8	1.3 - 1.5	4.4 - 4.8	1.30 - 1.35
petrifons	3.7 - 4.5	2.6 - 2.8	1.9 - 2.1	1.5 - 1.7	4.6 - 5.3	1.20 - 1.30
vanhyningi	3.0 - 3.5	2.1 - 2.5	1.6 - 1.8	1.3 - 1.5	4.4 - 4.8	1.35 - 1.50
parva	2.4 - 2.7	1.8 - 2.1	1.2 - 1.5	1.05 - 1.25	3.9 - 4.2	1.50 - 1.66
fraterna	2.8 - 3.6	1.9 - 2.4	1.3 - 1.7	1.1 - 1.4	4.0 - 4.8	1.30 - 1.45

differing from that of other species by having 2 or more elongate crests within the U-shaped superior tubercle.

C. petrifons (new species). Shell large with relatively large number of whorls per length. Whorls generally not as inflated. Lower corner of aperture weakly inflated where basal lip joins columellar margin. Operculum nearly multispiral with 3 large, slowly expanding whorls. Verge with only a single crest within the U-shaped superior tubercle.

C. wekiwae (new species). Shell small, thin, transparent. Suture deeply impressed. Peristome evenly rounded. Operculum paucispiral, with 2 rapidly expanding whorls. The verge is unique in having an invaginated cave-like pit within the U-shaped superior tubercle, and by having a slender constricted base.

C. parva (new species). Shell minute, spire short; suture weakly impressed. Operculum subspiral, consisting of about 2.5 regularly expanding whorls. Superior tubercles arranged in oblique longitudinal series. Accessory crest absent.

C. fraterna (new species). Shell small, translucent or nearly opaque. Suture weakly impressed. Peristome rounded. Operculum paucispiral, consisting of 2 rapidly expanding whorls. Verge with superior tubercles forming a vague, broken loop. Inferior crest very large and high. Accessory crest absent.

Cincinnatia vanhyningi (Vanatta)

Lyogyrus vanhyningi Vanatta, 1934, Naut., 47: 149; pl. 13, fig. 9.

Shell (Fig. 67, A). Conical, thick, translucent or opaque; light gray. Spire 0.8-1.0 times the length of the aperture, conical, very slightly convex. Umbilicus very narrowly rimate or imperforate. Sculpture consisting of fine incremental striations. 4.4-4.8 whorls; nuclear whorl protruding, about 0.30 mm in diameter; remaining whorls weakly arched. Suture weakly impressed, regularly descending to the last whorl, where it descends more rapidly to the aperture. Aperture ovate; inside of aperture hyaline, approaching white in very large specimens; upper corner angulate, outer lip forming an angle of about 90°. Peristome usually complete, but frequently incomplete across parietal wall. Outer lip thin, slightly arched forward in lateral profile; basal lip and columellar margin tending to form an angle at their intersection; lower corner slightly fluted and tending to extend as a slight tongue-like projection.

Length of shell, 3.0-3.5 mm; diameter of shell, 2.1-2.5 mm; length of aperture, 1.6-1.8 mm; width of aperture, 1.3-1.5 mm; length/diameter, 1.35-1.50; length/aperture, 1.80-2.00; whorls/length, 1.35-1.50.

Measurements of type (Vanatta, 1934: 149): length of shell, 2.4 mm; diameter of shell, 1.8 mm; length of aperture, 1.5 mm; width of aperture, 1.1 mm. The average size of the type and the paratypes is slightly less than that of the present population at the type locality.

OPERCULUM (Fig. 31, D). Ovate, thin, transparent, chitinous, amber colored. Nearly multispiral, with about 3 large whorls which slowly increase in size. Nucleus located in lower left third of operculum. Outer surface sparsely marked by fine incremental striations. Length of operculum, 1.20-1.45 mm; width of operculum, 0.90-1.10 mm.

VERGE (Fig. 45, E-H). The penis is relatively long and thick and extends considerably beyond the apical crest. Along its middorsal surface it bears a large glandular crest that extends nearly to the tip. A narrow glandular crest lies along the right margin. The superior tubercles are fused into a series of narrow crests. One series forms a ring that may be closed, or it may be broken into a series of segments or form a large U. The second series of crests lies about the periphery of the ring, and may be continuous with the narrow crest along the right margin of the penis. No crests or tubercles lie within the ring. The parapical crest is large, and is raised on a fleshy pedicel. A prominent inferior crest is present, but an accessory crest is lacking.

RADULA. Five specimens examined. The teeth are arranged in 46-58 transverse rows. The central tooth is 36-43μ wide; its cusp formula is $\frac{5\text{-}1\text{-}5}{1\text{-}1}$ or $\frac{6\text{-}1\text{-}6}{1\text{-}1}$. The lateral tooth has 2-1-3 or 2-1-4 cusps. The inner marginal tooth has 15-20 cusps.

TYPE LOCALITY. Seminole Springs, 3.4 miles northeast of Sorrento, Lake County, Florida. Type: *ANSP* 162084.

DISTRIBUTION (Fig. 20). Known only from the type locality.

ECOLOGY. The springs are located in two deep ravines, and are the headwaters of Seminole Creek, which flows into Black Creek, and this in turn flows into the Wekiva River. The springs emerge from caverns at the base of very steep banks, and form small pools. The run below the springs is fed by many small boils along its course. The bottoms of the run and pools are covered with gravel

FIGURE 20

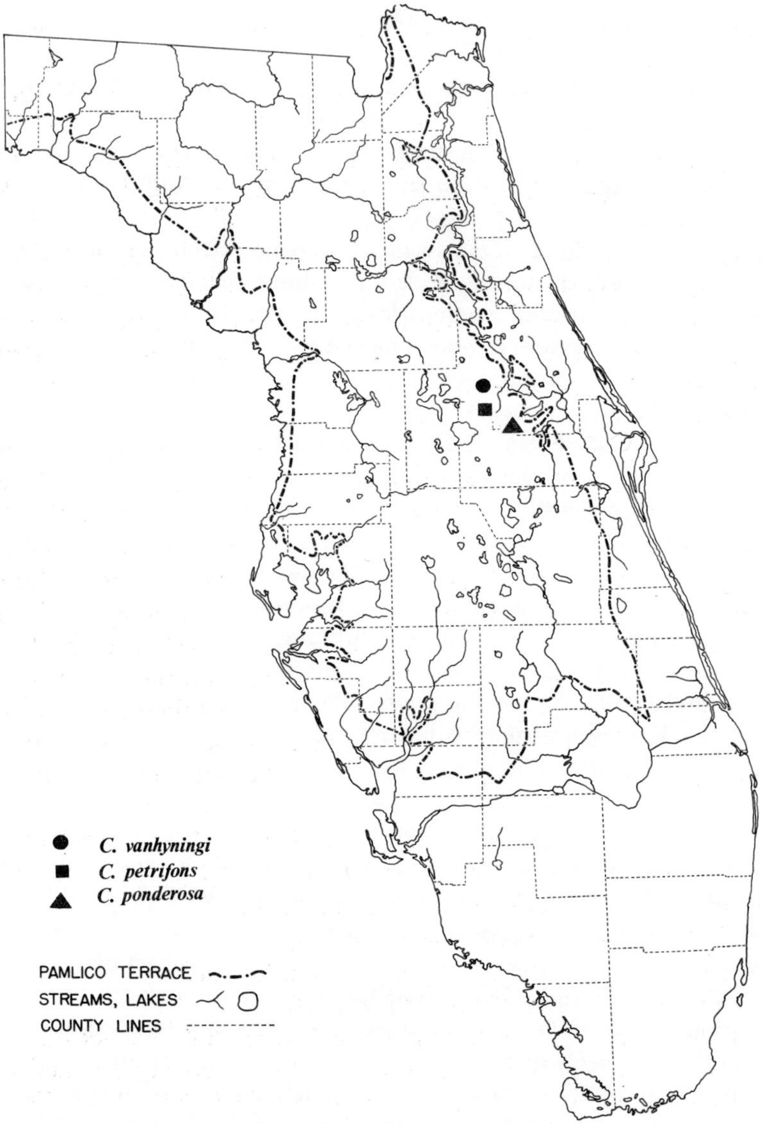

The geographic distribution of *Cincinnatia vanhyningi* (Vanatta), *Cincinnatia petrifons* new species, and *Cincinnatia ponderosa* new species.

and sand. The run and the pools contain large patches of eel grass and other vegetation. Snails were most abundant on the vegetation.

Cincinnatia ponderosa new species

SHELL (Fig. 67, C). Large, conical; thick, opaque; olivaceous. Very narrowly rimate or imperforate. Spire 0.85-1.00 times the length of the aperture, conical, slightly convex. Sculpture consisting of fine incremental striations. 4.3-4.9 whorls; nuclear whorl protruding, about 0.30 mm in diameter; early whorls rounded; body whorl evenly rounded, moderately shouldered. Suture deeply impressed; regularly descending to the last whorl where it descends more rapidly to the aperture. Aperture ovate; interior of aperture white; upper corner bluntly angulate or rounded. Peristome complete across parietal margin; outer lip relatively thick, straight, or slightly arched forward in lateral profile, nearly parallel to shell axis, evenly rounded with basal lip and columellar margin; columellar margin weakly concave; parietal margin straight, continuous, thick.

Length of shell, 4.0-4.6 mm; diameter of shell, 2.8-3.1 mm; length of aperture, 2.0-2.3 mm; width of aperture, 1.7-1.85 mm; length/diameter, 1.35-1.48; length/aperture, 1.85-2.00; whorls/length, 1.00-1.15.

Measurements of type: length of shell, 4.30 mm; diameter of shell, 2.95 mm; length of aperture, 2.10 mm; width of shell, 1.80 mm; 4.7 whorls; diameter of nuclear whorl, 0.30 mm.

OPERCULUM (Fig. 31, G). Ovate, transparent, chitinous, amber colored. Paucispiral, consisting of about 2 rapidly expanding whorls; nucleus located in lower left third of operculum. Outer surface sculptured with numerous fine incremental striations; inner surface smooth. Length of operculum, 1.75-2.00 mm; width of operculum, 1.30-1.50 mm.

VERGE (Fig. 46, A-C). The penis is moderately long, extending about as far as the apical crest of the verge. The superior tubercles form a series of crests that may be fused. One series forms a broad, open U which may occasionally be broken. Within the U lie 2 or 3 elongate crests. Additional small crests and tubercles may also be present within the U. An additional series of crests lies about the periphery of the U and extends onto the penis, as in *C. vanhyningi*. The parapical crest is raised on a thick pedicel. An accessory crest is present distal to the parapical crest. The inferior crest is very large and usually extends beyond the right margin of the verge.

RADULA. Five specimens examined. The teeth are arranged in 45-70 transverse rows. The central tooth is 43-48μ wide; its cusp formula is $\frac{6\text{-}1\text{-}6}{1\text{-}1}$ or $\frac{5\text{-}1\text{-}5}{1\text{-}1}$. The lateral tooth has 2-1-4 or 3-1-4 cusps. The inner marginal tooth has 14-18 cusps.

TYPE LOCALITY. Sanlando Springs, 3.1 miles west of Longwood, Seminole County, Florida. Type: *UMMZ* 216499; collected August 14, 1962 by Fred G. Thompson. Paratypes: *UMMZ* 216500 (500), *USNM* 636933 (120), *MCZ* 258013 (140), *ANSP* 301545 (30), *CM* 65-6 (100), *UF* (100).

DISTRIBUTION (Fig. 20). Known only from the type locality.

ECOLOGY. Sanlando Springs issues from a large limestone cavern about 15 feet below the water surface, and forms a large reniform pool about 50 feet across. The pool is situated at the base of some low rolling hills in a semi-wooded area along the east bank of the Little Wekiva River, and has been modified into a recreation area by damming the spring runs below the pool. Two runs flow from opposite ends of the pool, and these enter the Little Wekiva River about 200 yards apart. The bottoms of the pool, spring runs, and river are covered with sand and gravel. The runs and the river are filled with thick mats of aquatic vegetation, but the spring pool is frequently cleared to improve bathing conditions. The water throughout the system is clear and typically hard, containing about 100 ppm total hardness (Ferguson *et al.*, 1947: 152).

Cincinnatia ponderosa is abundant in the spring and in the Little Wekiva River to about 500 yards below the spring runs. It is not found beyond this point. Snails were most abundant on vegetation, although they were also common on sand and gravel.

Cincinnatia petrifons **new species**

SHELL (Fig. 67, D). Large, conical; thick, translucent or opaque; olivaceous. Very narrowly rimate or imperforate. Spire 0.85-1.20 times the length of the aperture, conical, straight-sided. Sculpture consisting of fine incremental striations. 4.6-5.3 whorls; nuclear whorl protruding, about 0.30 mm in diameter; whorls moderately rounded; body whorl evenly rounded, weakly shouldered. Suture moderately impressed, regularly descending to the aperture. Aperture ovate; interior of aperture white in mature shells; upper corner bluntly angulate, occasionally rounded; lower corner at juncture of basal lip and columellar margin tending to be angulate, weakly fluted internally and tending to project as a slight tongue-like extension of the lip. Peristome complete across parietal margin; outer lip evenly rounded to columellar margin, thick, nearly straight-edged in lateral profile, lying parallel with or at a slight angle to axis of shell; columellar margin weakly arched; parietal margin straight, callused.

Length of shell, 3.7-4.4 mm; diameter of shell, 2.6-2.8 mm; length of aperture, 1.9-2.2. mm; width of aperture, 1.5-1.7 mm; length/diameter, 1.37-1.50; length/aperture, 1.85-2.10; whorls/length, 1.20-1.30.

Measurements of type: length of shell, 4.20 mm; diameter of shell, 2.80 mm; length of aperture, 2.10 mm; width of aperture, 1.70 mm; diameter of nuclear whorl, 0.30 mm.

OPERCULUM (Fig. 31, F). Ovate, transparent, chitinous, amber colored. Nearly multispiral; consisting of about 3 large whorls that slowly increase in size; nucleus in the lower left third of the operculum. Outer surface smooth, shiny, with a few sparse incremental striations. Smooth inner surface. Length, 1.40-1.65 mm; width, 1.10-1.25 mm.

VERGE (Fig. 46, D-H). The penis is moderately long, extending to or slightly beyond the apical crest. The superior tubercles form a broad open U, within which lies a single long crest. This crest is rarely absent. Additional small tubercles may be present within the U, but hardly ever is a second crest developed. An additional series of crests lies about the periphery of the U, and may be continuous with the glandular crests on the penis. A small accessory crest and a very large inferior crest are present.

RADULA. Five specimens examined. The teeth are arranged in 40-66 transverse rows. The central tooth is 42-48μ wide; its cusp formula is $\frac{5\text{-}1\text{-}5}{1\text{-}1}$ or $\frac{6\text{-}1\text{-}6}{1\text{-}1}$. The lateral tooth has 2-1-3 cusps. The inner marginal tooth has 15-18 cusps.

TYPE LOCALITY. Rock Springs, 6.5 miles north of Apopka, Orange County, Florida. Type: *UMMZ* 216501; collected by Fred G. Thompson, November 29, 1963. Paratypes: *UMMZ* 216502 (500), *USNM* 636936 (90), *ANSP* 301536 (30), *MCZ* 258015 (130), *CM* 65-7 (60), *UF* (100).

DISTRIBUTION (Fig. 20). Known only from the type locality.

ECOLOGY. Rock Springs emerges from two horizontal caverns at the base of an overhanging limestone bluff. One source is from above the level of the spring run, and the second comes from a submerged cavern a few feet away from the first. These two springs form a fast-flowing spring run about 20 feet wide and 3 feet deep, which enters the Wekiva River a few miles away. The spring run is typically sand- and gravel-bottomed, and contains large patches of eel grass. Other aquatic plants are sparsely scattered along the shore and in quiet pools.

Snails are abundant and generally distributed from the boils at the origin of the run to about 300 yards downstream. Below this point they become very scarce and soon disappear. Snails are abundant on all forms of aquatic plants and on bare substrate. They are associated with *Notogillia wetherbyi, Pomacea paludosa,* and *Goniobasis vanhyningi.*

Cincinnatia wekiwae new species

SHELL (Fig. 67, B). Medium-sized; conical; thin, transparent; olivaceous. Very narrowly rimate or imperforate. Spire 0.90-1.05 times the length of the aperture, conical, straight-sided or slightly concave. Sculpture consisting of fine incremental striations. 4.4-5.0 whorls; nuclear whorl protruding, about 0.30 mm in diameter; remaining whorls moderately rounded; body whorl very weakly shouldered. Suture moderately impressed, descending regularly to the aperture. Aperture ovate; inside of aperture hyaline, occasionally with a very faint white deposit; upper corner bluntly angulate or rounded. Peristome usually complete, but very thin along parietal margin; outer lip nearly straight-edged in lateral profile, lying at an angle of 10° or less to the shell axis; peristome continuously rounded from outer lip to parietal margin; parietal margin straight, very thin near outer lip.

Length of shell, 3.1-3.7 mm; diameter of shell, 2.1-2.5 mm; length of aperture, 1.6-1.8 mm; width of aperture, 1.3-1.5 mm; length/diameter, 1.40-1.55; length/aperture, 1.90-2.05; whorls/length, 1.30-1.45.

Measurements of type: length of shell, 3.65 mm; diameter of shell, 2.55 mm; length of aperture, 1.85 mm; width of aperture, 1.50 mm; 5.0 whorls; diameter of nuclear whorl, 0.30 mm.

OPERCULUM (Fig. 31, H). Ovate, transparent, chitinous, amber colored. Paucispiral, with about 2 whorls which rapidly increase in size; nucleus located in the lower and left third of the operculum.

Outer surface sculptured with numerous fine incremental striations. Smooth inner surface. Length, 1.40-1.65 mm; width, 1.10-1.25 mm.

RADULA. Six specimens examined. The teeth are arranged in 39-56 transverse rows. The central tooth is 37-42μ wide; its cusp formula is $\frac{5\text{-}1\text{-}5}{1\text{-}1}$ or $\frac{6\text{-}1\text{-}6}{1\text{-}1}$. Lateral tooth with 2-1-3 or 2-1-4 cusps. Inner marginal tooth with 13-20 cusps.

VERGE (Fig. 47, A-D). The base is constricted and is proportionally slender. The penis is moderately long, extending slightly beyond the apical crest. A heavy glandular crest lies along its middorsal surface, extending nearly to the tip. The superior tubercles form a spiral arrangement of crests. Basically a U-shaped crest is formed, within which lies a long slender crest that may be continuous with the left limb of the U. Additional small tubercles may lie within the U. A single series of glandular crests lies along the periphery of the U and may extend onto the penis. Located within the U-shaped or spiral-shaped crest between the left limb of the U and the long median crest is a deep cave-like invagination, which extends beneath the left limb of the U-shaped crest. The parapical crest is conspicuously enlarged, sigmoid, and located on a thick pedicel. An accessory crest is present on the verge distal to the inner end of the parapical crest. The accessory crest is reduced in size and may occasionally be absent. An inferior crest is usually absent. When it is present it occurs as a small round tubercle.

Presumably the inferior crest is a vestigial structure that is in the process of being lost, rather than a structure that is making its phyletic appearance. This advanced degree of specialization in the verge is also indicated by the cave-like invagination of the outer surface.

TYPE LOCALITY. Wekiwa Springs, about 5 miles northeast of Apopka, Seminole County, Florida. Type: *UMMZ* 216503; collected August 22, 1963 by Fred G. Thompson. Paratypes: *UMMZ* 216504 (150), *USNM* 636938 (55).

DISTRIBUTION (Fig. 21). Known only from the type locality.

ECOLOGY. The springs originate from five submerged horizontal caverns, and form a pool about 120 feet across. At the time of my visits the main boil issued with such force that it formed a fountain shooting 1 to 2 feet above the surface of the pool. Collectively the springs discharge about forty-five million gallons of water per day. The bottom of the pool is uniformly covered with fine sand, as is also the spring run immediately below the pool. The spring run

FIGURE 21

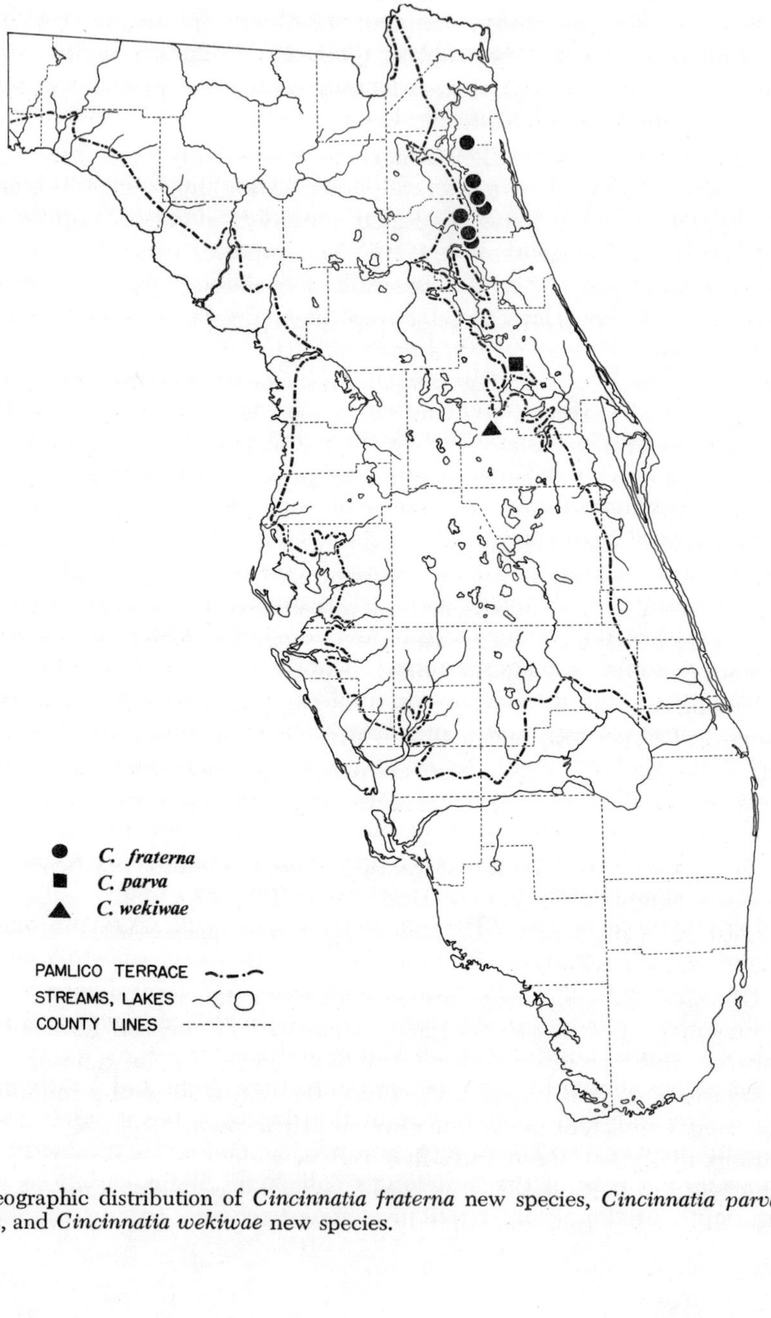

The geographic distribution of *Cincinnatia fraterna* new species, *Cincinnatia parva* new species, and *Cincinnatia wekiwae* new species.

is about 75 feet wide and 3 feet deep. The pool has a few small scattered clumps of plants, and algae coats the rocks lining the pool and the boils. The spring run contains thick mats of submerged vegetation, and large mats of hyacinths occur along the shores. The pool lies at the base of some low rolling sand hills. The spring run is bordered along both shores by low hardwood swamps, which continue along the Wekiva River.

Cincinnatia wekiwae occurs in the spring pool and the run for a short distance below the pool. It is abundant on aquatic vegetation, but is very scarce on the sand bottom.

Cincinnatia parva new species

SHELL (Fig. 66, F). Small, conical, 1.24-1.43 times as long as wide; moderately thin, transparent when clean. Spire conical, straight-sided or slightly convex, 0.78-1.08 times the height of the aperture but seldom exceeding the height of the aperture. Umbilicus narrowly rimate or imperforate. Periostracum olivaceous, usually covered with a thin rusty deposit. Surface of shell smooth or with a few fine incremental striations. 3.9-4.2 whorls; 1.50-1.66 whorls/length; nuclear whorl 0.23-0.28 mm in diameter, planular, protruding; remaining whorls weakly or moderately arched, only slightly curved between the suture and the periphery; whorls regularly descending to the aperture. Suture shallow or moderately impressed. Aperture ovate; upper corner bluntly angulate; interior hyaline; axis of aperture lying at about 35-40° to the axis of the shell; plane of aperture lying at about 15-20° to shell axis. Peristome continuous across parietal margin; adnate to preceding whorl; outer lip thin, without an internal callus, slightly sinuous in lateral profile, evenly rounded through columellar margin; parietal callus thin.

Length of shell, 2.40-2.70 mm; diameter of shell, 1.80-2.05 mm; height of aperture, 1.25-1.45 mm; width of aperture, 1.05-1.25 mm.

Measurements of the type: length of shell, 2.50 mm; diameter of shell, 1.80 mm; height of aperture, 1.30 mm; width of aperture, 1.10 mm; 4.1 whorls; nuclear whorl 0.23 mm in diameter.

Variations shown by the type series are very slight and mostly involve size and slight differences in the depth of the suture. The measurements given above represent the maximum size attained by about 99 per cent of the individuals collected. Slight variations in the depth of the suture result in some specimens having slightly more rounded whorls than usual. Other features of the shell are nearly constant in character.

OPERCULUM (Fig. 27, B). Ovate, apex bluntly pointed; chitinous, thin, transparent, amber in color. Paucispiral, nearly subspiral, consisting of about three and one-half whorls, the last rapidly expanding. Nucleus located at about one-third of the distance from the left margin to the right and from the base to the apex. Outer surface with very fine incremental striations. Inner surface smooth, with a small knob over the nucleus.

VERGE (Fig. 44, C-E). The verge is large and stout, and is attached to the body by a relatively slender neck. The penis extends to or beyond the tip of the relatively small apical crest. The parapical crest is thick and is slightly raised on a fleshy ridge. There are no accessory crests distal to the parapical crest. The inferior crest is medium-sized and is raised on a high fleshy pedicel. The superior tubercles are arranged in an oblique pattern, and one or two are always very long and slender. Two elongate tubercular crests are always present. One lies between the tip of the parapical crest and the middle of the opposite margin of the verge, and the other, which is about one-half as long as the first, lies between the first crest and the left margin. A small globular tubercle located near the base of the first elongate crest is usually present.

RADULA. The radula contains 55-63 transverse rows of teeth. The central tooth has 5 ectocones and a single basocone on each side of the mesocone. The lateral tooth has 2 entocones, a single mesocone, and 3-4 ectocones. The inner marginal tooth has 15-20 cusps, and the outer marginal tooth has about 20 very minute cusps.

TYPE LOCALITY. Blue Springs, 3 miles west of Orange City, Volusia County, Florida. Type: *UMMZ* 216530; collected April 21, 1962 by Fred G. Thompson. Paratypes: *UMMZ* 216531 (500), *ANSP* 301537 (30), *USNM* 636934 (100), *MCZ* 258014 (65), *CM* 65-10 (50), *UF* (50).

DISTRIBUTION (Fig. 21). Confined to the type locality.

ECOLOGY. The springs at the type locality form a large circular pool about 50 feet wide, and issue from a large central boil and several adjacent smaller boils. The spring run is about 30 yards wide and 5-15 feet deep; it is fed by many additional small boils along its course. The water in the spring pool and the run is very clear, and supports a thick luxurious growth of aquatic vegetation. The bottom of the pool and the run are covered with silty sands and gravel overlying a soft limestone. The spring run is about one-fourth of a mile long and enters the St. Johns River, which is very turbid and flows over a soft ooze bottom.

Cincinnatia parva is most abundant in the spring run near the spring, where it occurs on vegetation, debris, and gravel. I have not found it in the lower half of the spring run nor in the St. Johns River near the mouth of the run.

Cincinnatia fraterna new species

SHELL (Fig. 68, C). Medium-sized, conical, relatively narrow, 1.37-1.56 times as long as wide; thin, slightly transparent, translucent in very large specimens. Spire conical. straight-sided, relatively narrow, 0.93-1.28 times the height of the aperture. Umbilicus very narrowly rimate, occasionally imperforate. Periostracum light brown or olivaceous brown, with a dull luster. Sculpture consisting of distinct fine incremental striations that are usually most distinct near the umbilicus. These in turn may be crossed by very shallow subtle spiral striations that are usually most distinct near the suture. The spiral striations are not evident in many mature specimens. 4.0-4.8 whorls; 1.30-1.45 whorls/length; nuclear whorl 0.23-0.28 mm in diameter, protruding; remaining whorls regularly increasing in size, weakly arched. Suture shallow. Aperture subauriculate or ovate; upper corner angulate; interior hyaline, tinged with livid white; axis of aperture lying at about 25-35° to axis of shell; plane of aperture lying at about 10-15° to shell axis. Peristome complete across parietal margin at maturity, usually tightly adnate to the preceding whorl; outer lip thin, sharp, without an internal callus, slightly sinuous in lateral profile, evenly rounded to columellar margin; basocolumellar margin indented into umbilical region; parietal margin usually straight or weakly arched along curvature or preceding whorl, resulting in an auriculated aperture; parietal callus moderately thick at maturity.

Length of shell, 2.8-3.6 mm; diameter of shell, 1.9-2.4 mm; height of aperture, 1.3-1.7 mm; width of aperture, 1.1-1.4 mm.

Measurements of the type: length of shell, 3.25 mm; diameter of shell, 2.15 mm; height of aperture, 1.45 mm; width of aperture, 1.22 mm; 4.7 whorls; nuclear whorl 0.25 mm in diameter.

Little variation in shape and structure occurs within this species. Specimens from Fish Creek and from East Palatka, Putnam County, are slightly larger than normal and have a slightly more open umbilicus. No sexual dimorphism of shell characters occurs within this species.

OPERCULUM (Fig. 27, C). Ovate, apex bluntly pointed; chitinous, thin, transparent, light amber in color. Paucispiral, consisting of about 3 rapidly expanding whorls. The nucleus located at about one-third of the distance from the base to the apex, and at about two-fifths of the distance from the left to the right margin. Outer surface sculptured with a few fine incremental striations. Inner surface smooth.

VERGE (Fig. 44, F-H). The verge is very short, stout, and broad at the base and tapers distally. The penis does not extend beyond the sigmoid-shaped apical crest. The apical crest and the terminal bulb are reduced in size. The parapical crest is rather stout and is raised on a low fleshy crest. There are no accessory crests. The *inferior crest is very large and is raised on a very high fleshy pedicel.* The superior tubercles consist of a series of short, narrow, glandular crests that tend to be arranged in a semicircle. The lower half of the series may be missing or the series may coalesce into 1 or 2 long narrow crests. An additional elongate tubercle is usually present behind and below the tip of the parapical crest, and within the upper loop of the semicircular series of tubercles.

RADULA. The ribbon contains 45-52 transverse rows of teeth. The central tooth has 4-5 ectocones and a single basocone on each side of the mesocone. The lateral tooth has 2 entocones, a single large mesocone, and 3-4 ectocones. The inner marginal tooth has 16-19 cusps, and the outer marginal tooth has about 20 very small cusps.

TYPE LOCALITY. A small creek entering the St. Johns River, 4.0 miles south of Tocoi, St. Johns County, Florida. Type: *UMMZ* 216532; collected September 3, 1963 by Fred G. Thompson. Paratypes: *UMMZ* 216533 (100), *UMMZ* 216534 (50 in alcohol), *USNM* 636937 (40).

DISTRIBUTION (Fig. 21). This species is restricted to a small area in the northeastern part of the Florida peninsula, where it occurs in creeks, small streams, and sloughs along the lower third of the St. Johns River. The species is confined to the east side of the river, except for a small area in Putnam County near Palatka. Its distribution is nearly coincidental with that of *Aphaostracon rhadinus*.

LOCALITIES. Florida: *Putnam Co.*—Fish Creek, 3.3 mi. e. U.S. 17, 3.7 mi. s. Bostwick; bay along St. Johns River, East Palatka; creek,

East Palatka; creek 2 mi. n. East Palatka. *St. Johns Co.*—creek 0.7 mi. s. Tocoi; creek 6.5 mi. n. Tocoi; Trout Creek at Fla. 13 and Duval Co. line.

ECOLOGY. *Cincinnatia fraterna* occurs in streams that have a slow current, turbid water, and a sandy or sand-silt bottom. Specimens have been collected from submerged vegetation and the roots of floating plants, and from bottom sediments and debris. This species is commonly associated with *Littoridinops monroensis* and *Aphaostracon rhadinus*.

SUBFAMILY AMNICOLINAE

The subfamily Amnicolinae as defined by Morrison (1949, 13-15. = Bythinellinae) includes those species with a bifurcate verge that contains two ducts. One duct is a simple vas deferens that terminates at the tip of the penis. The other duct discharges through a flagellum on the left margin of the verge and orginates from a large convoluted blind duct in the coelom. The function of the duct is unknown. The operculum is chitinous and paucispiral, or some modification thereof.

Only a single North American genus is known to belong to this subfamily. The number of valid genera in Europe and Asia is still uncertain.

In most essential features this subfamily is very similar to the subfamily Buliminae. The only distinction between the two subfamilies lies in the structure of the operculum, which is made up of concentric rings in the Buliminae. When one considers the diverse opercular structures of the hydrobiid genera that inhabit Lake Titicaca which clearly are relatively recently derived from *Heleobia*, the opercular differences between the Amnicolinae and the Buliminae are not very impressive.

Amnicola Gould, 1841

Amnicola Gould, 1841, Rept. Invert. Mass.: 228-29.—Stimpson, 1865, Smithsonian Misc. Collections, (201): 12-17.—Berry, 1943, Misc. Publ. Museum Zool. Univ. Mich., (57): 22-23.—Pilsbry, 1943, Naut., 57: 69.—Baker, 1947, Naut., 60: 105-06.—Baker, 1960, Bull. Zool. Nom., 18: 148. Type species: *Paludina porata* Say, 1821 =

Amnicola limosa var. *porata* (designated by Herrmannsen, 1846: 38).

Lyogyrus Gill, 1863, Proc. Acad. Nat. Sci. Phila.: 34.—Pilsbry, 1892, Naut., 6: 61.—Pilsbry, 1916, Naut., 30: 84.—Walker, 1918, Misc. Publ. Museum Zool. Univ. Mich., (6): 33, 147. Type by original designation: *Valvata pupoidea* Gould, 1841.

Euamnicola Crosse and Fischer, 1891, Miss. Sci. Mex. Moll., 2: 261.—Pilsbry, 1943, Naut., 57: 69.—Baker, 1947, Naut., 60: 105. Type species: *Paludina porata* Say, 1821 = *Amnicola limosa* var. *porata;* objective synonym of *Amnicola* Gould.

Marstoniopsis Altena, 1936, Basteria, 1: 68.—Morrison, 1947, Naut., 60: 86-87. Type by original designation: *Hydrobia steinii* Martens, 1858.

Shell minute to medium-sized; variable in shape, globose to elongate-conical; thin and transparent to thick and opaque. Umbilical perforation one-sixth to one-tenth the diameter of the shell, not conspicuously obscured by reflected columellar lip. Spire conical, straight-sided or convex in outline, 0.6-1.8 times the length of the aperture. Periostracum dull in texture; sculpture consisting of incremental striations which are usually very fine but may be coarse. 3-5 whorls at maturity. Nuclear whorl 0.29-0.48 mm in diameter, depressed or rounded and protruding. Remaining whorls inflated, may be shouldered. Suture deeply impressed. Aperture ovate or rhomboid; interior hyaline or livid white; axis of aperture at about 25-40° to axis of shell; plane of aperture at about 10-30° to shell axis. Peristome continuous or discontinuous across parietal wall at maturity; outer lip sharp-edged; basal lip and columellar lip of variable curvature, not conspicuously thickened.

Operculum thin, membranous, transparent, colorless or light amber yellow, ovate or nearly circular. Paucispiral to nearly spiral with 3.0-4.5 whorls. Outer surface dull with a few incremental striations. Inner surface smooth, glassy, with a small knob over the nucleus.

The foot is broad and spatulate, truncate anteriorly and rounded posteriorly. Operculigerous lobes not extended beyond the margins of the foot. Tentacles about one and one-half times as long as the snout when extended. Eyes located at the outer base of the tentacles. Snout, mantle, and tentacles variously mottled with dark pigment. The gill originates near the middle of the mantle collar and extends posteriorly and to the left to the anterior left corner of the pericardium. The gill consists of about 19-33 triangular lamellae which are most strongly developed along the middle or the anterior half of the series.

The verge originates on the nape at a distance behind the tentacles about equal to the length of the snout and slightly to the right so that its left margin lies on the middorsal line. When erect the verge extends anteriorly over the right side of the body. When contracted it is recurved beneath the mantle with the end of the penis arched around the end of or lying beneath the flagellum and pointing posteriorly. The verge is bifid. Its left margin bears a robust flagellum that is flat-topped and is nearly equal to the length of the penis. A slender duct winds down through the flagellum, passes back along the left margin of the verge, and enters the coelom where its expands into a large saculate convoluted blind duct that lies in a coiled middorsal mass behind the tentacles. At the point where the duct discharges through the flagellum it produces a relatively large circular open pore. The penis is cylindrical in shape and may be stocky or slender. The vas deferens enters the base of the verge and passes along the right margin to the tip of the penis.

No conspicuous sexual dimorphism occurs, except that the females tend to be more robust than the males. The female is oviparous and lays large eggs that bear a wide chitinous ring that circumscribes the area of attachment, and a similar wide chitinous crest that passes across the top of the egg capsule.

The radula is typically taenioglossate with about 50-97 transverse rows of teeth. The central tooth is trapezoidal with 1-4 basocones on each side of the face and 4-6 ectocones on each side of a large mesocone. The central tooth is about 20-40μ wide. The lateral tooth is large and clavate with 1-4 entocones, a large mesocone, and 3-6 ectocones. The inner marginal tooth is broad and spatulate and has about 20-30 cusps. The outer marginal tooth is long and slender with a flange-like lateral expansion along the basal half of the tooth. Its rasping edge bears about 20-35 cusps, which are too minute to be clearly distinguished under oil immersion magnification.

The genus *Amnicola* is widely distributed in North America east of the Rocky Mountains and in Europe. Two subgenera are recognized in North America, *Amnicola* s.s. and *Lyogyrus*. *Amnicola* is distinguished from *Lyogyrus* by its generally larger shell with a large nuclear whorl, its stout verge, and its strongly mottled mantle. *Lyogyrus* has a minute shell with a small nuclear whorl; its verge is relatively slender, and its mantle is diffusely pigmented with melanophores.

Amnicola has been used as a catch-all for most American amnicoloid species that could not conveniently be placed elsewhere on the basis of their shells, and has been thought to include many species from as far south as Panama. The Central American species that have been placed in *Amnicola* belong in other genera as follows. These changes are based upon examination of the anatomy of most of the species. The species will be discussed elsewhere.

Aroapyrgus Baker, 1931

cisternina Martens
conchensensis Walker
costaricensis Morch
guatemalensis Crosse and Fischer
hinkleyi Walker
panamensis Tryon
panzosensis Walker
pasionensis Goodrich and van der Schalie
petenensis Morelet
stolli Martens
subangulatus Martens
tryoni Pilsbry

Pyrgophorus

hyalina Morelet
hydrobioides Ancey
volcani Martens

Zetekina

melanioides Martens
tenuis Martens

The other American species, except those discussed by Berry (1943, 23-41), and Taylor (1966, 173), and those discussed in this paper, remain of undetermined generic status.

Subgenus *Amnicola* Gould, 1841

The shell is generally medium-sized with a large nuclear whorl, 0.38-0.48 mm in diameter. The penis and the flagellum are relatively stout, and the mantle is heavily mottled with black.

Amnicola s.s. is known to include only the type species *A. limosa* and its varieties, *A. dalli* (Pilsbry and Beecher) and its subspecies *johnsoni* Pilsbry, and *A. rhombostoma* n. sp. It probably also includes *A. parva* Lea and *A. pallida* Haldeman. The distinctions between *limosa, dalli, parva,* and *pallida* are subject to doubt, and they may eventually be found to be variations of a single widely distributed species.

Amnicola s.s. is confined to North America east of the Rocky Mountains. Three forms are known to occur in Florida: A. *d. dalli*, *A. d. johnsoni*, and *A. rhombostoma*. Berry (1943, 23) stated that *A. limosa* also occurs in Florida, but did not cite any records. I have not been able to identify any of the Florida populations with *A. limosa*. It is distinguished from the Florida forms by its larger size and its peristome, which is complete across the parietal wall at maturity.

The three forms of *Amnicola* that occur in Florida may be distinguished as follows.

Amnicola d. dalli—Shell medium-sized, 2.2-4.0 mm long, conical or globose-conical; thin, translucent; umbilical perforation one-tenth or less the diameter of the shell; incremental striations uniformly weak; 3.1-4.0 whorls; nuclear whorl 0.38-0.48 mm in diameter; aperture ovate; peristome discontinuous across parietal wall; basal lip not indented posteriorly; aperture 1.3-2.2 mm long; operculum multispiral.

Amnicola d. johnsoni—Shell medium-sized, 2.2-3.7 mm long, conical or globose-conical, thin, translucent or opaque; umbilical perforation one-eighth to one-tenth the diameter of the shell; incremental striations uniformly weak; 3.0-3.9 whorls; nuclear whorl 0.41-0.48 mm in diameter; aperture ovate; peristome discontinuous across parietal wall; basal lip not indented posteriorly; length of aperture 1.3-2.0 mm; operculum paucispiral.

Amnicola rhombostoma—Shell moderately large, 3.5-4.8 mm long, conical, moderately thick, opaque; umbilical perforation one-sixth to one-eighth the diameter of the shell; incremental striations intensified near suture and umbilicus; 3.8-4.3 whorls; nuclear whorl 0.43-0.48 mm in diameter; aperture rhomboid; basocolumellar angle extended as a gouge-like projection; peristome discontinuous across parietal wall; basal lip broadly but shallowly indented; length of aperture 2.0-2.5 mm; operculum paucispiral.

Amnicola d. johnsoni and *A. d. dalli* were proposed as species belonging to different genera (Pilsbry, 1899: 22). The similarities of shell structure between the two forms were noted by Pilsbry, but he recognized them as species of different genera because of differences of their opercula that he regarded as reliable generic characters. *Dalli* is usually distinguished from *johnsoni* by possessing a loosely coiled multispiral operculum as opposed to a relatively more tightly coiled paucispiral operculum. No other significant differences occur between the two forms. Many population samples of *Amnicola*, particularly from springs along the Suwannee River, have shells that are referable to either form, and the opercula are intermediate in structure. All degrees of variation occur so that it becomes a subjective matter to assign many individuals to either nominate form.

Since the two forms tend to be allopatric in their distributions, are ecologically separated, and intergrade in all conchological and opercular characters, I consider them to be subspecifically related. Apparently *dalli* is a derivation of the ecologically and geographically more widely distributed *johnsoni*.

The resulting combination of names is an excellent example of why patronymics should generally be avoided in taxonomy.

Amnicola dalli johnsoni Pilsbry

Amnicola johnsoni Pilsbry, 1899, Naut., 13: 2.—Walker, 1918, Misc. Publ. Museum Zool. Univ. Mich., (6): 134.

SHELL (Fig. 69, A, B; Fig. 55, D-F). Globose-conical, 1.05-1.30 times as long as wide; translucent or opaque, moderately thick, particularly around the columellar area; periostracum light tan or light brown with a dull luster. Umbilical perforation narrow, about one-eighth to one-tenth the diameter of the shell, not conspicuously obscured by reflected columella. Spire globose-conical, 0.61-0.97 times height of aperture; strongly convex in outline, 0.6-0.9 times the length of the aperture. Sculpture nearly smooth, consisting of fine, inconspicuous incremental striations that are of nearly uniform intensity over the shell. 3.1-3.9 whorls; nuclear whorl depressed, planorboid, 0.41-0.48 mm in diameter; remaining whorls rapidly increasing in size, globose, weakly or moderately shouldered. Suture deeply impressed, descending to the aperture at an increasing rate. Aperture broadly ovate or rhomboid; upper corner of aperture bluntly angulate or rounded; interior of aperture hyaline or slightly livid-white due to a slight thickening of the lip; axis of aperture at about 30-45° to the axis of the shell; plane of aperture at about 10-15° to axis of shell. Peristome sharp, may be slightly thickened internally, usually discontinuous across parietal wall; outer lip evenly rounded to basocolumellar angle, which extends slightly anteriorly and basally; outer lip very slightly arched forward in lateral profile; columellar margin indented into umbilical region, slightly reflected.

Measurements: length, 2.2-3.7 mm; diameter, 2.1-3.4 mm; length of aperture, 1.3-2.0 mm; width of aperture, 1.2-1.8 mm; length/aperture, 1.61-1.97; whorls/mm, 1.03-1.28.

This form shows considerable variation in size, obesity, and the shape of the aperture. These variations are not correlated with any ecological or geographical factors, although each population tends to have its own characteristic shell size and shape.

OPERCULUM (Fig. 28, E). Operculum medium-sized, ovate, hyaline, light amber in color, thin, membranous, paucispiral, consisting of about three and three-quarter large whorls, the outer one expanding more rapidly than the others. Nucleus located slightly to the left side of the midline, and at about two-fifths of the distance from the base to the apex. Outer surface smooth with a few fine radial striations. Inner surface glassy, smooth, with a small round knob over the nucleus.

ANIMAL (Fig. 27, G). The tentacles bear two dark stripes that extend from the tip to the base. The stripe on the inner surface may be very light or diffuse. The outer stripe is usually dark, and continuous with the pigmentation on the side of the body. The interorbital area bears a large, ragged V-shaped blotch. The nape is unpigmented, as is the tip of the snout. The top and sides of the snout may be gray in color. The sides of the body have a broad, irregular, black band that extends posteriorly from the base of the tentacles to the mantle, and may spread over the sides of the pedicel. The inner surface of the mantle collar is light gray. The outer surface of the collar and a narrow zone of the mantle posterior to it is unpigmented. The mantle bears a narrow, irregular, transverse black band that lies parallel to the collar. An irregular and sometimes diffuse blotch extends along the dorsal surface of the spire, overlying the uterus, intestine, and liver. Another narrow black blotch extends posteriorly from near the middle of the collar to the left margin of the pericardium and partially overlies the gill. Much of the area over the gill is generally unmarked as is a small area over the terminal end of the intestine posterior to the transverse band.

The gill contains about 22-33 triangular lamellae, which are largest near the middle of the series. The lamellae taper slightly in size anteriorly, and become very much reduced posteriorly.

VERGE (Fig. 37, D, G). When erect the verge is about 2.5 mm long. When contracted it is recurved to the left under the mantle, with the tip of the penis arched over the end of the flagellum and pointing posteriorly. The verge is slightly compressed dorsoventrally so that it is oval in cross section. It is large and sickle-shaped with a large flat-topped flagellum on the left margin. A very slender duct winds through the flagellum, passes posteriorly along the left side of the base of the verge, and terminates as an enlarged and highly convoluted blind pouch which lies in the coelom at the base of the verge slightly to the right of the middorsal line. The vas deferens is relatively large and lies along the right side of the verge. It becomes much reduced in size at the point where it enters the body.

RADULA. The radula contains about 65-71 transverse rows of teeth. The central tooth is about 32-36μ wide and about one-half as high. It bears 2-3 basocones and 4-6 ectocones on each side of the face. The lateral tooth is large and clavate and has 1-3 entocones, a large mesocone, and 3-4 ectocones. The inner marginal tooth has about 23-30 cusps on the rasping margin. The outer marginal tooth has about 20 cusps on the rasping edge.

TYPE LOCALITY: St. Augustine, St. Johns County, Florida. Lectotype by present designation: *ANSP* 58089.

DISTRIBUTION (Fig. 22). This species is known only from Florida where it is found throughout the northern two-thirds of the peninsula from Highland County and Manatee County north to St. Johns County and Suwannee County. It has also been taken at one locality in Leon County near Tallahassee.

LOCALITIES: Florida: *Alachua Co.*—Sante Fe R., 3 mi. n.w. High Springs. *Bradford Co.*—Heilbronn Springs, 5.6 mi. n.n.w. Starke. *Citrus Co.*—Crystal River; Homosassa Springs; Withlacoochee R. below dam. *Columbia Co.*—Santa Fe R., O'Leno State Park. *DeSoto Co.*—Joshua Creek, 0.6 mi. n. Nocatee; creek 1.6 mi. s. Fort Ogden; creek 6.6 mi. s.s.w. x-sec. Fla. 760 and Fla. 761; Joshua Creek, 3.0 mi. s.e. Arcadia; slough 2.5 mi. s.w. Fort Ogden; Peach R., 1.9 mi. w.n.w. Fort Ogden; creek 0.7 mi. s.s.w. Brownville; Horse Creek, 8.0 mi. w.s.w. Arcadia. *Hardee Co.*—creek 3.6 mi. e. Ona; Charlie Creek, 8.4 mi. s. Zolfo Springs; Horse Creek, 6.5 mi. w. Ona; Peace R., 1.0 mi. e. Wauchula; Payne Creek, 1.5 mi. s. Bowling Green. *Hillsborough Co.*—Lithia Springs, 3.0 mi. w.n.w. Lithia; Purity Springs, Tampa; Little Manatee R., 6.3 mi. e. Wimauma; Lake Thonotasassa; Hillsborough R. at U.S. 41; Hillsborough R., Hillsborough River State Park; Sweetwater Creek, 7.5

FIGURE 22

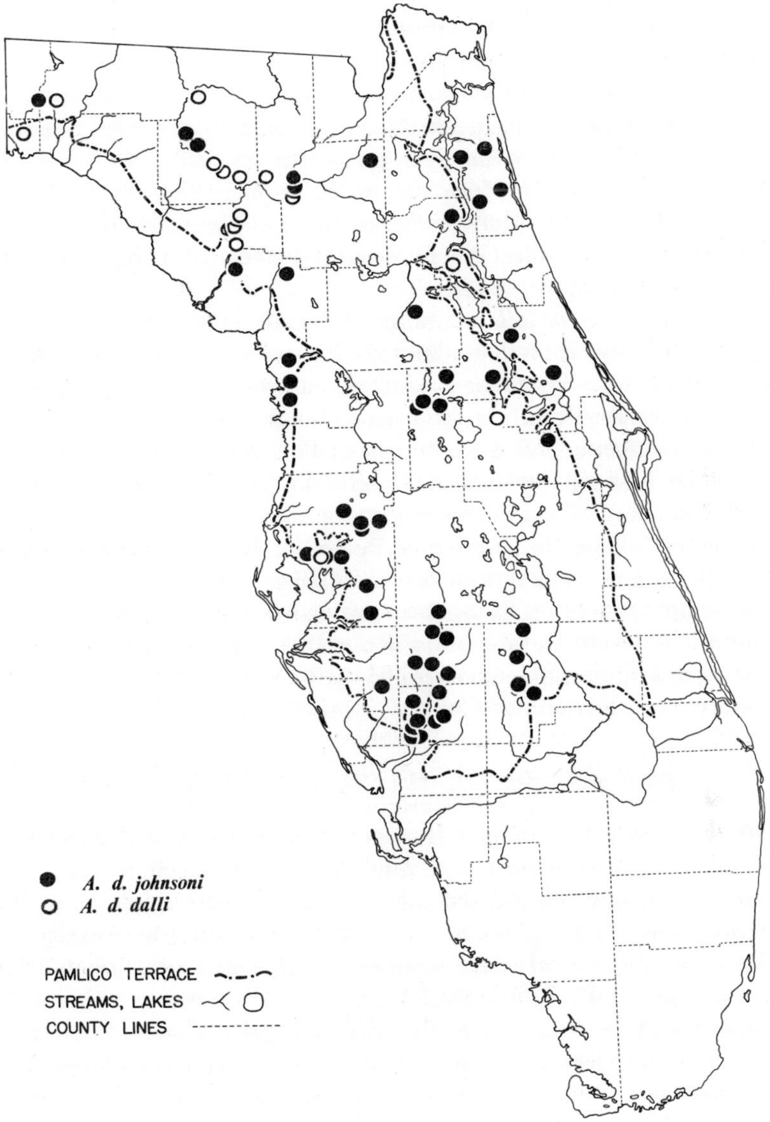

The geographic distribution of *Amnicola d. dalli* (Pilsbry) and *Amnicola d. johnsoni* Pilsbry.

mi. n.w. Tampa. *Highlands Co.*—Lake Ann, 1.0 mi. s.e. Lake Placid (City); Arbuckle Creek, 9.0 mi. e.n.e. Avon Park; Josephine Creek at U.S. 27; Lake June-in-Winter. *Lake Co.*—Blackwater Creek, 2.0 mi. s.w. Cassia; west end Lake Harris, 2.0 mi. s. Leesburg; s.e. end Lake Harris at Fla. 19; Bugg Springs, 0.3 mi. n.w. Okahumpka; Lake Yale, 3 mi. n.w. Grand Island. *Leon Co.*—Natural Bridge Springs. *Levy Co.*—Blue Springs, 3 mi. w. Bronson; Manatee Springs, 6.0 mi. w. Hardeetown. *Manatee Co.*—Myakka R., 0.5 mi. e.s.e. Myakka City. *Marion Co.*—creek on Fla. 314, 6.0 mi. n.e. Grahamsville. *Orange Co.*—Econlockhatchee R. 3.0 mi. w. Bithlo. *Pasco Co.*—Crystal Springs; small creek 1.0 mi. n.n.e. Wesley Chapel. *Polk Co.*—creek 3.2 mi. s. Fort Meade. *Putnam Co.*—Welaka Springs, 1.0 mi. n. Welaka; slough, East Palatka; Fish Creek, 3.7 mi. s. Bostwick. *St. Johns Co.*—creek 3.1 mi. n.n.w. Spuds; creek 2.3 mi. s. Switzerland; creek 0.9 mi. n.n.w. Durbin; creek 2.9 mi. s.w. St. Augustine. *Suwannee Co.*—Charles Springs, 5.0 mi. n.w. Luraville; Telford Springs, Luraville. *Volusia Co.*—Bay Creek 5.0 mi. e. Osteen.

ECOLOGY. This form is generally found in small sand-bottomed streams. Occasionally it occurs in larger sand-bottomed rivers, lakes, and springs. It is usually associated with streams having a slow or moderate current and clear water. Aquatic vegetation was sparse at most stations, and specimens were found on the substrate or on tree rootlets.

Amnicola dalli dalli (Pilsbry and Beecher)

Lyogyrus dalli Pilsbry and Beecher, 1892; Naut., 6: 62.—Walker, 1918; Misc. Publ. Museum Zool. Univ. Mich., (6): 147.

SHELL (Fig. 69, C, D; Fig. 55, A-C). Medium-sized, conical or globose-conical, 1.04-1.33 times as long as wide; occasionally globose; thin, fragile, subtransparent or translucent; periostracum light gray or light brown, dull. Umbilical perforation narrow, about one-tenth or less the diameter of the shell, partially obscured by the reflected columellar margin of the lip. Spire conical, straight-sided, or convex in outline, 0.6-0.9 times the length of the aperture. Sculpture indistinct, consisting of very fine, irregular incremental striations that are nearly uniform in intensity over the shell. 3.1-4.0 whorls; nuclear whorl flat-topped, depressed, 0.38-0.48 mm in diameter; remaining whorls rapidly increasing in size, strongly inflated but not shouldered. Suture deeply impressed, usually descending to the aperture at an increasing rate. Aperture ovate or nearly circular; upper corner obtusely angulate; interior hyaline; axis of aperture at about 30° to axis of shell; plane of aperture at about 15-20° to shell axis. Peristome discontinuous across parietal margin; outer lip thin, sharp, fragile, very slightly arched forward in lateral profile, nearly straight-edged; basocolumellar margin slightly thickened and may tend to be weakly angulate; columellar margin indented into umbilical region, moderately reflected.

Length, 2.2-4.0 mm; diameter, 2.1-3.4 mm; length of aperture, 1.3-2.2 mm; width of aperture, 1.2-2.0 mm; length/aperture, 1.59-1.95; whorls/mm, 1.00-1.42.

There is considerable variation in the size and shape of the shell. The variation within a single population may be nearly as great as that between any two populations. Most colonies contain specimens that have a short globose spire as well as specimens that have a longer conical spire. Usually the shape of the aperture is ovate, but occasional populations are dominated by individuals with a circular opening. These variations are not correlated with any ecologic or geographic factors.

Operculum (Fig. 28, B). Typically small, broadly ovate, hyaline, slightly amber in color, thin, membranous, nearly spiral, consisting of about 4 slowly expanding whorls. Nucleus located on the midline and at about two-fifths of the distance from the base to the apex. Outer surface smooth, but with a few fine radial striations. Inner surface glassy, smooth, with a small knob over the nucleus.

The nearly spiral nature of the operculum is variable among some population samples. The rate of expansion of the last whorl and the location of the nucleus is not consistent, and occasionally the operculum approaches the paucispiral condition of *johnsoni*. The only structural difference in the two types of opercula is in the rate of expansion of the last whorl. Some population samples, particularly from springs along the Suwannee River, are impossible to assign with certainty to either subspecies because all degrees of intergradation occur in opercular structure as well as in shell characters.

Animal. The animal is like *johnsoni* except for some differences in the color patterns. In life the stripe on the inner side of the tentacles consists predominantly of bright fluorescent yellow flecks, which are lost in preservation, leaving only the pattern of the melanophores. In addition the pattern on the mantle of *dalli* is generally not as intense or as clearly delimited as it is in *johnsoni.*

VERGE (Fig. 37, E). The verge is similar in structure, arrangement, and size to that of *johnsoni.* When erect it is about 2.0-2.5 mm long.

RADULA (Fig. 36, B). The ribbon contains about 63-70 transverse rows of teeth. The central tooth is about 40-43μ wide and has 2-3 basocones and 4-5 ectocones on each side of the face. The lateral tooth has a single entocone, a large mesocone, and 3-4 ectocones. The inner marginal tooth has about 25-30 cusps. The outer marginal tooth has about 20 cusps.

DISTRIBUTION (Fig. 22). This subspecies is known only from Florida where it occurs throughout the northern half of the peninsula and into the panhandle at least as far west as Leon County.

LOCALITIES. Florida: *Alachua Co.*—Hornsby Springs, 3.0 mi. n. High Springs. *Columbia Co.*—Ichetucknee Springs, Mill Spring. *Dixie Co.*—spring 2.0 mi. s. Rock Bluff Ferry. *Gilchrist Co.*—Bell Springs, 0.8 mi. n.w. Fanning Springs; Otter Springs, 3 mi. w.s.w. Wilcox; Hart Springs; Rock Bluff Springs. *Hillsborough Co.*—Sulphur Springs, Sulphur Springs Park. *Jefferson Co.*—Wacissa Springs, Wacissa. *Lafayette Co.*—Troy Springs, 6 mi. n.w. Branford; spring 1.4 mi. s.e. Troy Springs. *Orange Co.*—Wekiwa Springs. *Suwannee Co.*—Branford Springs, Branford; Falmouth Springs, Falmouth. *Volusia Co.*—Benson's Mineral Spring, Enterprise (*USNM*); Ponce De Leon Springs. *Wakulla Co.*—St. Marks R., Newport.

TYPE LOCALITY. Wekiva River, Seminole County, Florida. (This is the Wekiva River that flows from Wekiwa Springs, Orange County, and not the Wekiva River that flows from Wekiva Springs, Levy County.) Lectotype by present designation: *ANSP* 64611.

ECOLOGY. This subspecies is confined to springs, spring runs, and spring-fed streams. In such situations the water is very clear and is relatively constant in temperature throughout the year. Specimens were always found in thick clumps of aquatic vegetation, particularly *Elodea* and *Vallisneria,* and were almost always abundant.

Amnicola rhombostoma new species

SHELL (Fig. 69, E, F; Fig. 55, G, H). Moderately large, stout, conical, 1.09-1.33 times as long as wide; moderately thick; opaque or only slightly translucent; periostracum olive brown or rusty brown, with a dull oily luster. Umbilical perforation moderately wide, about one-sixth to one-eighth the diameter of the shell, partially obscured by the reflected columella. Spire conical, slightly convex, 0.6-1.0 times the length of the aperture. Sculpture consisting of fine, irregular, incremental striations on the base and on the periphery of the whorls. The striations become more distinct in the umbilicus and on the shoulders of the whorls near the suture. 3.8-4.3 whorls; nuclear whorl flat, slightly elevated, 0.43-0.48 mm in diameter; remaining whorls regularly increasing in size, inflated, moderately shouldered. Suture deeply impressed, regularly descending to the aperture. Aperture rhomboid, occasionally broadly ellipsoid; upper corner bluntly rounded; basocolumellar corner bluntly angulate or rounded, slightly extended anteriorly and basally as a gouge-like protrusion; interior of aperture livid white; axis of aperture at about 25-30° to the axis of the shell; plane of aperture at about 30° to shell axis. Peristome discontinuous across parietal margin; outer lip and basal lip sharp-edged, frequently with a slight callus deposit within; outer lip arched slightly forward in lateral profile; basal margin with a broad shallow indentation posteriorly, which is accentuated by the basocolumellar projection; basocolumellar corner of shell conspicuously thickened; columellar margin angulate, indented into umbilical region, slightly reflected.

Length, 3.5-4.8 mm; diameter, 3.0-3.9 mm; length of aperture, 2.0-2.5 mm; width of aperture, 1.7-2.1 mm; length/aperture, 1.59-2.00; whorls/mm, 0.90-1.11.

Measurements of type: length, 4.35 mm; diameter, 3.75 mm; length of aperture, 2.50 mm; width of aperture, 2.10 mm; 4.1 whorls; nuclear whorl 0.43 mm in diameter.

Most variations of the shell involve the size and shape of the aperture, but even these characters are nearly constant in nature.

OPERCULUM (Fig. 28, A). Large, broadly ovate, hyaline, nearly colorless, thin, membranous, paucispiral, consisting of about three and one-half large, slowly expanding whorls. Nucleus located slightly to the left side of the midline, and at about two-fifths of the distance from the base to the apex. Sculpture consisting of a few distinct radial striations. Outer surface smooth; inner surface smooth, glassy, with a small knob over the nucleus.

ANIMAL. Each tentacle bears two dark bands that extend from the tip to the base. The band on the outer surface is more intense than that on the inner side. The interorbital area bears a posteriorly pointing ragged V-shaped blotch. The nape and the tip of the muzzle are unpigmented. The rest of the snout posterior to the interorbital blotch is dark smokey-gray on the top and the sides. The sides of the body bear a broad black band that is continuous with the outer band on the tentacles and the interorbital blotch and extends posteriorly to the mantle. This band may also spread out over the sides of the pedicel. The inner surface of the mantle collar is light gray in color. The outer surface of the collar is unmarked. The mantle has a narrow transverse black blotch that lies immediately behind the collar. A broad, irregular, and sometimes diffuse blotch extends along the spire overlying the uterus, intestine, and part of the liver and is continuous with the collar blotch. Another narrow and darker blotch extends posteriorly along the left margin of the gill. The region overlying the gill and a small area overlying the terminal portion of the intestine posterior to the transverse band are unpigmented.

The gill contains about 26-31 triangular lamellae. The lamellae are largest near the collar and become progressively smaller posteriorly.

VERGE (Fig. 37, F). Similar to that of *A. d. johnsoni* in shape and arrangement. In large males it is about 2.5 mm long when erect.

RADULA (Fig. 36, A). As in *A. d. johnsoni* except that the radula contains about 50-60 transverse rows of teeth. The central tooth is 44-57μ wide, has 3-4 basocones and 4-5 ectocones on each side. The lateral tooth has 2-3 entocones, a large mesocone, and 3-4 ectocones. The inner marginal tooth has about 20-26 cusps. The outer marginal tooth has about 25-30 cusps.

TYPE LOCALITY. Peters Creek at Florida State Highway 209, 3.5 miles north-northwest of Green Cove Springs, Clay County, Florida. Type: *UMMZ* 216505; collected March 21, 1963 by Fred G. Thompson. Paratypes: *UMMZ* 216506 (30); same data as the type.

DISTRIBUTION (Fig. 23). This species is confined to small sand-bottomed streams and rivers draining into the west side of the St.

FIGURE 23

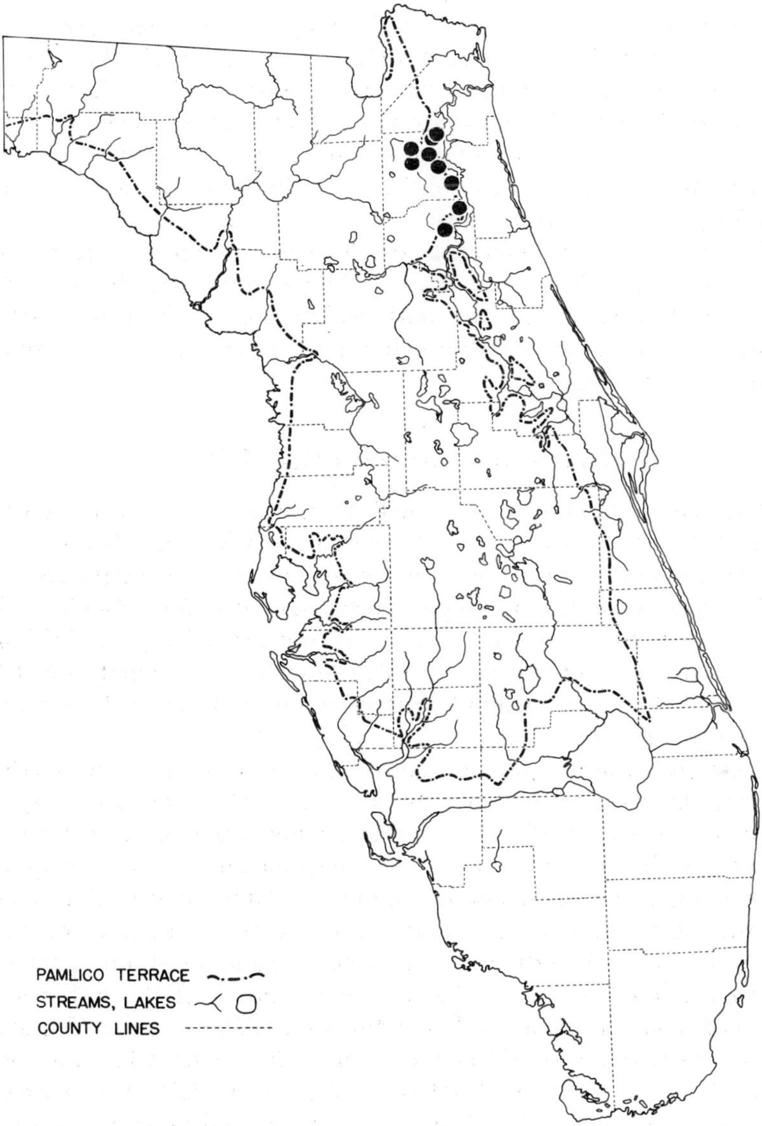

The geographic distribution of *Amnicola rhombostoma* new species.

Johns River in Clay and Putnam Counties, Florida. It is sympatric and syntopic with *Amnicola retromargo.*

LOCALITIES. Florida: *Clay Co.*—North Fork of Black Creek, 14 mi. s.w. Orange Park; stream 1.5 mi. s.w. Orange Park; Wadeboro Springs, 2 mi. s.s.w. Orange Park; South Fork of Black Creek, 0.2 mi. s. Middleburg; Black Creek at Fla. 209, 1 mi. s. x-sec. Fla. 220; Clark Creek at Fla. 209, 9.1 mi. n. Bostwick. *Putnam Co.*—stream on Fla. 209, 0.2 mi. e. U.S. 17, 3.7 mi. s. Bostwick; Rice Creek on Fla. 100, 7.3 mi. w.n.w. Palatka.

ECOLOGY. As with *Amnicola retromargo,* specimens were found in streams and rivers that have a slow current and drain sandy soils forested with pine and hardwood scrub. Snails were usually found on the sandy substrate, but also occurred on submerged vegetation and debris.

Subgenus *Lyogyrus* Gill, 1863

The shell is minute with a small nuclear whorl 0.29-0.36 mm in diameter. The penis and the flagellum are relatively slender and elongate, and the mantle is diffusely shaded with melanophores.

Lyogyrus includes the type species *A. pupoidea* (Gould), *A. browni* Carpenter, *A. grana* (Say), *A. walkeri* Pilsbry, *A. pilsbryi* Walker, and *A. retromargo* n.sp. The subgenus is confined to North America, where it is generally distributed east of and including the Mississippi River system.

Previous authors have recognized *Lyogyrus* as a genus distinguished by the multispiral operculum that characterizes its type species. Pilsbry (1916, 83) established the subfamily Lyogyrinae, which he distinguished from the Amnicolinae by its multispiral operculum, an action generally accepted by later authors. The value of the operculum as a taxonomic character is limited in this group of snails. The difference between a paucispiral and a multispiral condition is due to the rate of expansion of the last whorl (*A. pupoidea, A. grana,* and *A. retromargo,* Fig. 28, C, D, F), and the multispiral nature of the operculum is inconsistent in some species (*A. d. dalli* and *A. d. johnsoni,* Fig. 28, B, E). Also, species that have no close relationship to *Lyogyrus* would be placed with it because of opercular characteristics (i.e., *Cincinnatia vanhyningi*).

The species of *Lyogyrus* are alike in having minute shells with very small nuclear whorls, and are anatomically very similar. The anatomy of four species of *Lyogyrus* has been examined. Berry (1943, 28) reported on the anatomy of *A. walkeri* from Michigan, and I have examined animals of *A. pupoidea* and *A. grana* from North Carolina, and *A. retromargo* from Florida (Fig. 37, A-C). The reproductive structures of these four species are alike, even in minor details. The color patterns of the mantles of *A. pupoidea* and *A. grana* are similar to that described below for *A. retromargo*. Berry omitted this character in his discussion of *A. walkeri*.

Because of similarities of conchological characters and soft anatomy, the six species listed above appear to comprise a natural, closely interrelated group. The group is very similar to *Amnicola s.s.*, and does not deserve more than subgeneric recognition. The operculum is useful only as a specific criterion and has no significance above that level.

Heterocylus Crosse, 1872, from New Caledonia has been considered to be subgenerically related to *Lyogyrus* (Franc, 1957: 44; Solem, 1961: 429), but there is no basis for this other than the unreliable similarities of their opercula.

The species of *Lyogyrus* may be separated into two groups on the basis of shell shape. This is a matter of convenience and does not necessarily reflect biological relationships. *Amnicola pupoidea* and *A. browni* differ from other species by being more elongate with spires that are one and one-half times the length of the aperture or greater. *Amnicola grana, A. walkeri, A. pilsbryi,* and *A. retromargo* are alike in being more broadly conical, with spires that seldom exceed one and one-quarter times the length of the aperture. These four species are very much alike in size and general appearance. *Amnicola retromargo* differs from the others by its broadly indented basal lip and its forward projecting columellar lip. There is a tendency for these characters also to occur in some specimens of *A. walkeri,* but never to the extent that they do in *A. retromargo*. *Amnicola walkeri* is further distinguished from *A. retromargo,* as well as from others, by having a smaller aperture and a light tan or gray periostracum. The periostraca of the other species are darker rusty brown. *Amnicola pilsbryi* is unique in having an incomplete peristome across the parietal margin of the aperture at maturity. *Amnicola grana* is characterized by having a tightly coiled multispiral operculum. In this respect it is similar to *A. pupoidea* and *A. browni.*

Amnicola (Lyogyrus) retromargo new species

SHELL (Fig. 68, A, B; Fig. 56, A-F). Small, conical, 1.05-1.29 times as long as wide; thin, semitransparent; periostracum light rusty brown or light gray-brown, with a dull luster. Umbilical perforation open, about one-sixth to one-eighth the diameter of the shell, not conspicuously obscured by the reflected columella. Spire conical, 0.74-1.09 times length of aperture; straight-sided or only slightly convex. Sculpture consisting of very fine but distinct incremental striations that are nearly equal in intensity over the shell. 3.8-4.3 whorls in shells with a complete peristome; nuclear whorl 0.29-0.33 mm in diameter, rounded, protruding; remaining whorls regularly increasing in size, inflated, may be moderately shouldered. Suture deeply impressed, regularly descending to the aperture. Aperture ovate or nearly so; upper corner of aperture rounded; interior hyaline or slightly tinged with white; axis of aperture at about a 35-40° angle to axis of the shell; plane of aperture at about a 30° angle to shell axis. Peristome continuous across parietal wall at maturity; outer lip thin, sharp, slightly arched forward in lateral profile, very slightly thickened within; basal lip broadly but shallowly indented posteriorly; columellar margin moderately indented into umbilical region, usually rounded but may be angulate; lower part of columellar margin with a small forward projection that is accentuated by the indented basal margin.

Length, 2.0-2.8 mm; diameter, 1.9-2.3 mm; length of aperture, 1.1-1.4 mm; width of aperture, 1.0-1.25 mm; length/diameter, 1.05-1.29; length/aperture, 1.74-2.09; 1.50-1.91 whorls/mm.

Measurements of type: length, 2.50 mm; diameter, 2.15 mm; length of aperture, 1.30 mm; width of aperture, 1.10 mm; 4.0 whorls; diameter of nuclear whorl, 0.33 mm.

This species is conservative in most shell features. Most variations involve only size and relative width of the shell. Specimens from Peters Creek, Clay County are unusually large and relatively narrow, but show no other noticeable differences from specimens of other populations.

OPERCULUM (Fig. 28, C). Operculum small, broadly ovate, hyaline, colorless, thin, membranous; paucispiral, consisting of about three and one-half rapidly expanding whorls. Nucleus located about one-third of the distance from the left margin to the right, and about two-fifths of the distance from the base to the apex.

Outer surface smooth, with a few very fine radial striations. Inner surface glassy, with a small knob over the nucleus.

ANIMAL (Fig. 27, H). The inner sides of the tentacles bear a heavy black band that extends from the tip to the base. The outer surfaces of the tentacles lack bands, unlike other Florida species of *Amnicola.* A large dark interorbital blotch lies at the base of the tentacles. Usually the snout is unpigmented. The sides of the body have a broad black band that extends from the base of the tentacles posteriorly to the mantle, and continues with the interorbital blotch by passing around the front of the tentacles. The rest of the nape, pedicel, and foot are unmarked. The mantle bears a diffuse, dark, broad transverse band that lies behind the collar and may extend posteriorly over the gill to the pericardium. The rest of the spire may be diffusely mottled or reticulated with melanophores.

The gill extends diagonally from near the middle of the mantle collar to the anterior left margin of the pericardium. It consists of 19-29 triangular lamellae that are largest slightly anterior to the middle of the series and become progressively smaller posteriorly.

VERGE (Fig. 37, C). The erect verge is about 1.5 mm long. When contracted it is recurved to the left side under the mantle with the tip of the penis arched over the tip of the flagellum and pointing posteriorly. The verge is slightly compressed dorsoventrally so that it is oval in cross section. The penis and the flagellum are relatively long and slender, as they are in *A. walkeri* (Berry, 1943: 28), *A. pupoidea,* and *A. grana* (Fig. 37, A, B), in contrast to the more robust condition in *A. limosa, A. dalli,* and *A. rhombostoma.*

RADULA (Fig. 36, C). The ribbon contains about 90-97 transverse rows of relatively small teeth. The central tooth is about 27-29μ wide and is about one-half as high. It bears a single basocone and 4-5 ectocones on each side of the face. The lateral tooth is hatchet-shaped, and has 3-5 small entocones, a large mesocone, and 4-5 small ectocones. The inner marginal tooth has about 24-26 cusps. The outer marginal tooth has about 25 very small cusps.

TYPE LOCALITY. Rice Creek at Florida State Highway 100, 7.3 miles west-northwest of Palatka, Putnam County, Florida. Type: *UMMZ* 216507; collected March 21, 1963 by Fred G. Thompson. Paratypes: *UMMZ* 216508 (46).

DISTRIBUTION (Fig. 24). The species is known only from a narrow zone across the neck of the Florida peninsula from the west

FIGURE 24

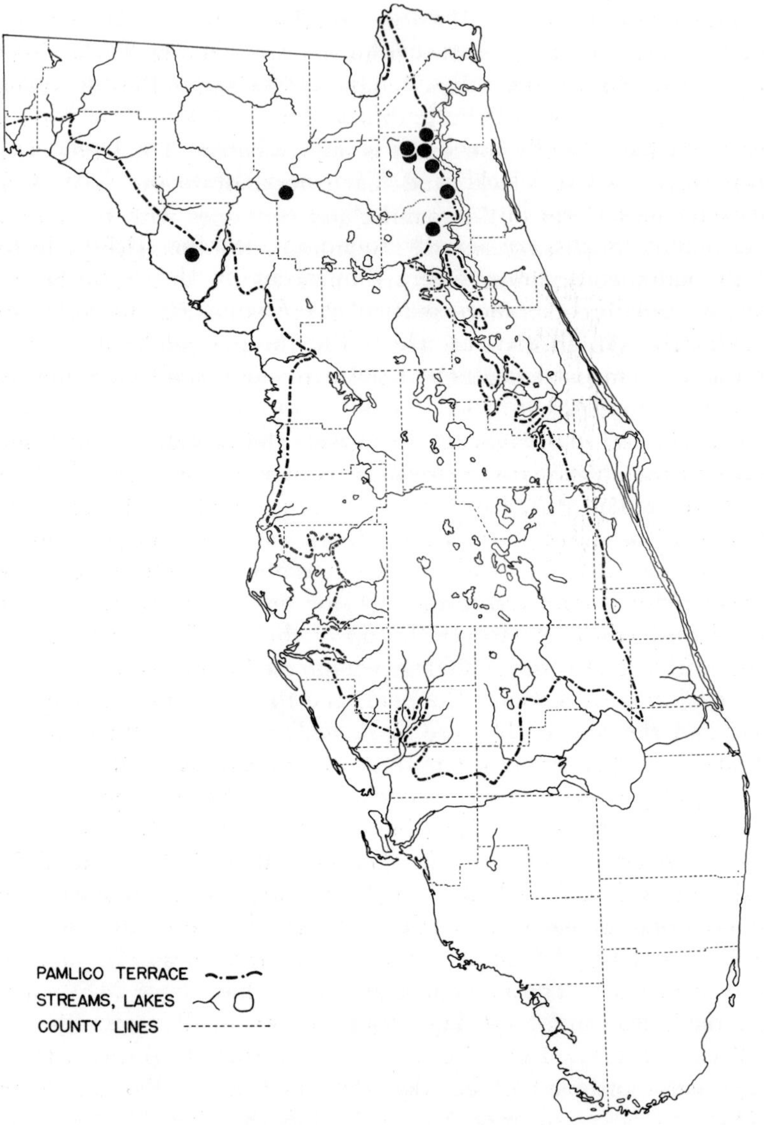

The geographic distribution of *Amnicola retromargo* new species.

side of the St. Johns River in Clay and Putnam Counties west to Dixie County.

LOCALITIES. Florida: *Alachua Co.*—Santa Fe R., 3.0 mi. n.w. High Springs. *Clay Co.*—stream 1.5 mi. s.w. Orange Park; North Fork Black Creek, 14 mi. s.w. Orange Park; South Fork Black Creek, 0.2 mi. s. Middleburg; Black Creek at Fla. 209, 1 mi. s. x-sec. Fla. 220; Peters Creek at Fla. 209, 3.5 mi. n.n.w. Green Cove Springs; Clark Creek on Fla. 209, 9.1 mi. n. Bostwick. *Dixie Co.*—creek 4.5 mi. s. Cross City.

ECOLOGY. Snails are usually found in clear or slightly turbid sandy-bottomed streams, where they occur on the substrate, debris, or aquatic vegetation, particularly *Elodea, Hydrocotyla,* and *Eichhornia.* At all stations where they were found, the streams had a slow current and drained through sandy soil that supported growths of pine or hardwood deciduous forests. Aquatic vegetation was generally sparse except for occasional clumps of hyacinths (*Eichhornia*). *Amnicola retromargo* is syntopic with *A. rhombostoma* where they occur together.

Literature Cited

ABBOTT, R. T., and H. S. LADD. 1951. A new brackish-water gastropod from Texas (Amnicolidae: *Littoridina*). J. Wash. Acad. Sci., 41: 335-38.

ANCEY, C. F. 1888. Etude monographique sur le genre *Pyrgulopsis.* Bull. Soc. Malac. de France, V: 185-202.

BAKER, F. C. 1928. Freshwater Mollusca of Wisconsin; Pt. I. Gastropoda. Bull. Wisconsin Geol. Nat. Hist. Surv., 70(1): i-xx, 1-539; pls. 1-58.

BAKER, H. B. 1924. Land and freshwater mollusks of the Dutch Leeward Islands. Occasional Papers Museum Zool. Univ. Mich., (152): 1-158; 21 pls.

______. 1930. The Mollusca collected by the University of Michigan-Williamson Expedition in Venezuela. Occasional Papers Museum Zool. Univ. Mich. (210): 1-92; 12 pls.

BARTSCH, P. 1953. A new Rissoid Mollusk from Maryland. Naut., 67: 40-41; pl. 5, fig. 5.

BERRY, E. G. 1943. The Amnicolidae of Michigan: distribution, ecology and taxonomy. Misc. Publ. Museum Zool. Univ. Mich., (57): 1-68.

CLENCH, W. J. 1938. Land and freshwater mollusks of Grand Bahama and the Abaco Islands, Bahama Islands. Mem. Sociedad Cubana Historia Natural, 12: 303-33.

______. 1963. Land and freshwater mollusks of the Crooked Island group, Bahamas. Bull. Museum Comp. Zool. Harvard Coll., 128(8): 395-413.

CLENCH, W. J., and R. D. TURNER. 1956. Freshwater mollusks of Alabama, Georgia, and Florida from the Escambia to the Suwannee River. Bull. Florida State Museum, 1(3): 97-239.

CONNOLLY, M. 1939. A monographic survey of South African non-marine Mollusca. Ann. S. African Museum, 33: 1-660.

COOKE, C. W. 1939. Scenery of Florida interpreted by a geologist. Florida Geol. Surv., Geol. Bull., 17: 1-118; 53 figs.

______. 1945. Geology of Florida. Florida Geol. Surv., Geol. Bull., (29): 1-339.

DALL, W. H. 1885. Notes on some Floridian land and freshwater shells. Proc. U.S. Natl. Museum, 8: 255-89.

______. 1890-1903. Contribution to the tertiary fauna of Florida. Wagner Free Institute of Science. Trans., 3(2): 1-1654.

______. 1910. A new Floridian *Amnicola.* Naut., 24: 2.

______. 1913. On a brackish water Pliocene fauna of the Southern Coastal Plain. Proc. U.S. Natl. Museum, 46: 225-37; 3 pls.

DOERING, A. 1884. Apuntes sobre la fauna de Moluscos de la República Argentina. Bol. Acad. Natl. Cienc. (Cordoba, Rep. Arg.), VII: 457-74.

EMILIANAI, C. 1957. Oxygen isotopes and paleotemperature determinations. Acts Fourth Intern. Cong. Quart. Rome, 1953. 2: 831-44.

FERGUSON, G. E., C. W. LINGHAM, S. K. LOVE, and R. O. VERNONE. 1947. The springs of Florida. Florida Geol. Surv., Geol. Bull., (31): 1-196.

FRANC, A. 1957. Mollusques terrestres et fluviatiles de l'archipel Néo-Caledonien. Mem. Museum Natl. Hist. (Paris), Ser. A, Zool., 13: 1-200; pls. 1-24.

FRAUENFELD, G. R. 1863. Vorlaufige Aufzahlung der Arten der Gattungen *Hydrobia* Atm. und *Amnicola* Gld.-Hald. in der Kaiserlichen und in Cumings Samlung. Verhandl. Zool. Botan. Ges. in Wein, 1017-32.

GOODRICH, C. 1944. Certain operculates of the Coosa River system. Naut., 58: 1-10.

HAAS, F. 1955. The Percy Sladen Trust Expedition to Lake Titicaca. XVII. Mollusca: Gastropoda. Trans. Linn. Soc. (London), 1(3): 275-308.

HUBENDICK, B. 1955. The Percy Sladen Trust Expedition to Lake Titicaca in 1937. The anatomy of the Gastropoda. Trans. Linn. Soc. (London), 1(3): 309-27; 95 figs.

HUBRICHT, L. 1957. New species of *Fontigens* from Shenandoah National Park. Naut., 71: 9-10.

———. 1963. New species of Hydrobiidae. Naut., 76: 138-40.

———. 1963. *Notogillia wetherbyi* in Alabama. Naut., 76: 152.

MORRISON, J. P. E. 1939. Notes on the genera *Potamopyrgus* and *Lyrodes*. Naut., 52(3): 87-88.

———. 1946. The non-marine mollusks of San Jose Island, with notes on those of Pedro Gonzales Island, Pearl Islands, Panama. Smithsonian Inst. Misc. Collection, 106(6): 1-49; 3 pls.

———. 1949. The cave snails of eastern North America. News Bull. Ann. Rept. Am. Malac. Union, 1948: 13-15.

———. 1965. New brackish water mollusks from Louisiana. Proc. Biol. Soc. Wash., 78: 217-24; figs. 1-9.

MUUS, B. J. 1963. Some Danish Hydrobiidae with the description of a new species, *Hydrobia neglecta*. Proc. Malac Soc. London, 35: 131-38.

NEILL, W. F. 1957. Historical biogeography of present-day Florida. Bull. Florida State Museum Biol. Sci., 2(70): 175-220.

PARODIZ, J. J. 1957. New records of fresh-water gastropods from the Bahama Islands. Ann. Carnegie Museum, 35: 1-9.

PILSBRY, H. A. 1889. New and little known American mollusks. No. 1. Proc. Acad. Nat. Sci. Phila., 81-89.

———. 1890. Notices of new Amnicolidae. Naut., 4: 63-64.

———. 1891. Land and fresh-water mollusks collected in Yucatan and Mexico. Proc. Acad. Nat. Sci. Phila., 310-34.

———. 1892. On the Amnicoloid genus *Lyogyrus*, with the description of a new species. Naut., 6: 61-62.

———. 1899. New Amnicolidae from Florida. Naut., 13: 20-22.

———. 1904. A new Florida *Amnicola*. Naut., 17: 113-14.

———. 1911. Non-marine Mollusca of Patagonia. Rep. Princeton Univ. Exp. to Patagonia: 1896-1899. 3: 513-633; pls. 38-47.

———. 1916. Notes on *Valvata micra* Pils. and Ferr. Naut., 30: 83-87.

———. 1930. List of land and fresh-water mollusks collected on Andros, Bahamas. Proc. Acad. Nat. Sci. Phila., 82; 297-302; 22 pls.

———. 1934. Pliocene freshwater fossils of the Kettleman Hills and neighboring California oil fields. Naut., 48: 15-17.

———. 1935. Mollusks of the freshwater Pliocene beds of the Kettleman Hills and neighboring oil fields, California. Proc. Acad. Nat. Sci. Phila., 86: 541-70; 6 pls.

———. 1950. New fountain snails from Florida. Naut., 64: 37-39.

———. 1951. New species of *Amnicola* and *Streptostyla*. Naut., 64: 119-20.

———. 1952. *Littoridina tenuipes* (Couper). Naut., 66: 50-54.

———. 1953. Pliocene Mollusca of southern Florida, with special reference to those from North Saint Petersburg: Vitrinellidae and freshwater Mollusks. Monogr. Acad. Nat. Sci. Phila., (8): 411-47.

PILSBY, H. A., and J. BEQUAERT. 1927. The aquatic mollusks of the Belgian Congo, with a geographical and ecological account of Congo malacology. Bull. Am. Museum Nat. Hist., 53: 69-602.

RHOADS, S. N. 1899. Annotated list of land and freshwater shells recently collected in the vicinity of Miami, Florida. Naut., 13: 43-48.

RUSSELL, R. S. 1957. Instability of sea level. Am. Scientist, 45: 414-30.

SCHUCHERT, C. 1935. Historical geology of the Antillean-Caribbean region. New York: i-xxvi, 1-811; 107 figs., 16 maps.

SOLEM, A. 1961. Hydrobiid snails from Lake Pontchartrain, Louisiana. Naut., 74: 157-60.

———. 1961. New Caledonian land and freshwater snails, and annotated check list. Fieldiana, Zool., 41(3): 419-501.

SOULEYET. 1852. Voyage autour du monde exécuté pendant les années 1836 et 1837 sur la corvette La Bonita, Zoologie, II. La Bonita, Zool., II: 565.

STIMPSON, W. 1865. Researches upon the Hydrobiinae and allied forms. Smithsonian Misc. Collections, 1-59.

TAYLOR, D. W. 1960. Late Cenozoic Molluscan faunas from the high plains. U.S. Geol. Surv. Prof. Pap., (337): i-iv, 1-94; pls. 1-4.

———. 1966. A remarkable snail fauna from Coahuila, Mexico. Veliger, 9: 152-228.

THIELE, J. 1928. Revision des systems der Hydrobiiden und Melaniiden. Zool. Jahr., 55: 351-402; 1 pl.

———. 1931. Handbuch der systematischen Weichtierkunde, Erster Band. Stuttgart, 1931. i-vi, 1-778.

TROSCHEL, F. H. 1856-93. Das Gebiss der Schnecken zur Begründung einer naturlichen Classification. Berlin (Nicolaische Verl.-Buckhandl.), 2 vols.

VANATTA, E. G. 1934. *Lyogyrus vanhyningi* new species. Naut., 47: 149; 1 pl., 1 fig.

WALKER, B. 1918. A synopsis of the classification of the freshwater Mollusca of North America, north of Mexico, and a catalogue of the more recently described species, with notes. Misc. Pub. Museum Zool. Univ. Mich., (6): 1-213.

WALTERS, V., and C. R. ROBINS. 1961. A new toadfish (Batrachoididae) considered to be a glacial relict in the West Indies. Am. Museum Nov., (2047): 1-24.

WENZ, W. 1924. Zur Nomenklatur Tertiärer Land und Süsswassergastropoden VI. Senckenbergiana, 6: 221-23.

Illustrations

FIGURE 25

Animals of various Florida hydrobiids

A. *Pyrgophorus platyrachis* new species. Female. Canal, Ochopee, Collier County, Florida.

B. *Pyrgophorus platyrachis* new species. Male. Canal, Ochopee, Collier County, Florida.

C. *Hyalopyrgus aequicostatus* (Pilsbry). Female from a shell with 6.1 whorls. Lake Panasoffkee, Sumter County, Florida.

D. *Hyalopyrgus aequicostatus* (Pilsbry). Male from a shell with 4.8 whorls. Blue Springs, 3 mi. w. Orange, Volusia County, Florida.

E. *Littoridinops monroensis* (Frauenfeld). Male, dorsal view. Quarry 3 mi. w. North Miami Beach, Dade County, Florida.

F. *Littoridinops tenuipes* (Couper). Male, lateral view. Ditch entering St. Marys River, near U.S. 17, Nassau County, Florida.

G. *Littoridinops monroensis* (Frauenfeld). Lateral view of specimen E.

Each scale equals 1 mm.

FIGURE 25

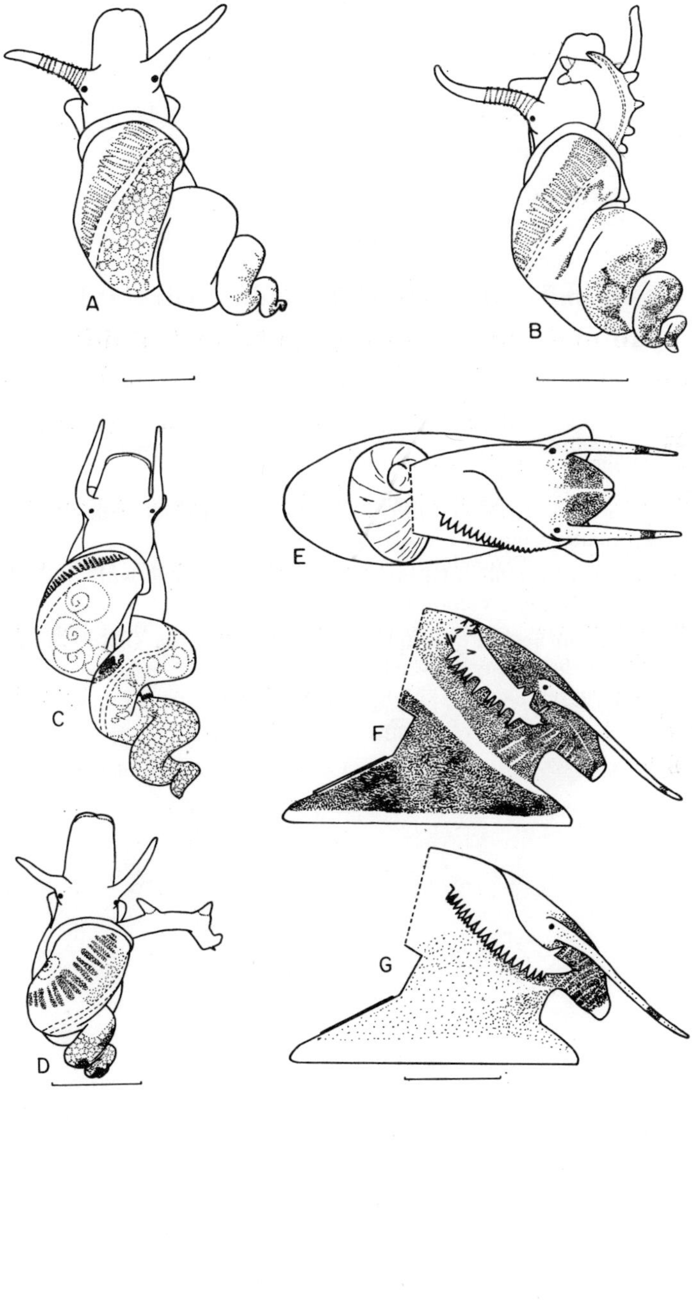

FIGURE 26

Animals of various Florida hydrobiids

A. *Notogillia wetherbyi* (Dall). Female, dorsal view. Lake Eustis, Lake County, Florida.

B. *Notogillia wetherbyi* (Dall). Ventral view of specimen A.

C. *Spilochlamys conica* new species. Male, dorsal view. Manatee Springs, Levy County, Florida.

D. *Cincinnatia floridana* (Frauenfeld). Male, dorsal view. Lake, Ponte Vedra Beach, St. Johns County, Florida.

E. *Heleobops docima* new species. Female. Pond 2.5 mi. n.e. West Lake, Everglades National Park, Dade County, Florida.

F. *Heleobops docima* new species. Male. Pond 2.5 mi. n.e. West Lake, Everglades National Park, Dade County, Florida.

Each scale equals 1 mm.

FIGURE 26

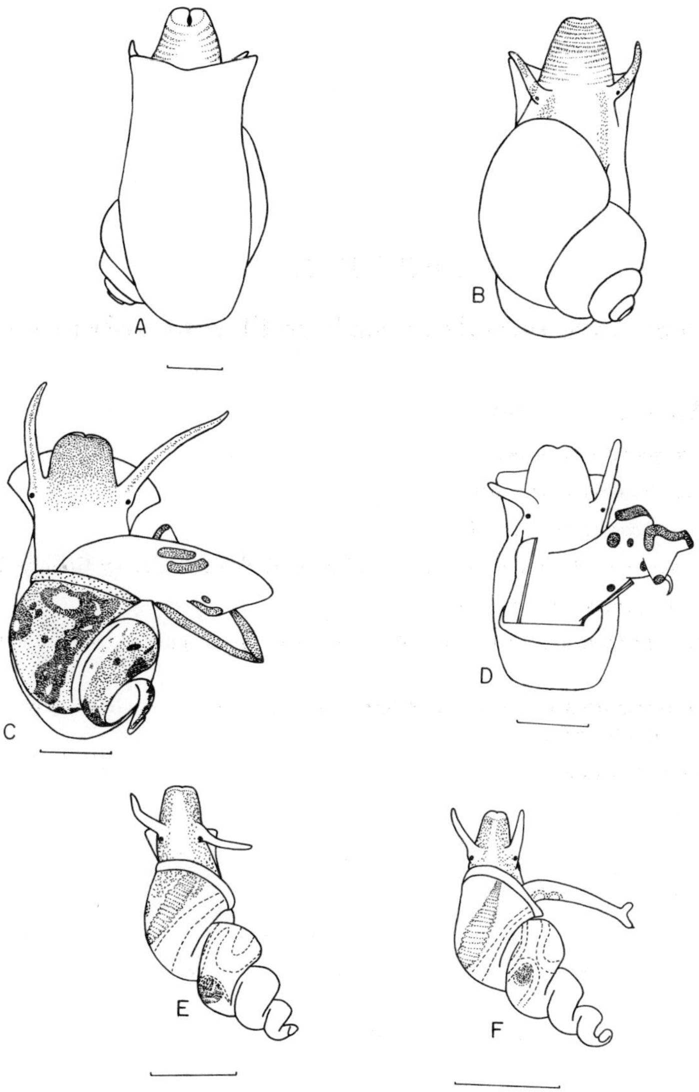

FIGURE 27

Opercula and animals of various Florida hydrobiids

A. *Cincinnatia monroensis* (Dall).

B. *Cincinnatia parva* new species.

C. *Cincinnatia fraterna* new species.

D. *Spilochlamys conica* new species.

E. *Onobops jacksoni* (Bartsch). Marsh 3.2 mi. s.w. Yankeetown, Levy County, Florida.

F. *Spilochlamys gravis* new species.

G. *Amnicola dalli johnsoni* Pilsbry. Creek 2.9 mi. s.w. St. Augustine, St. Johns County, Florida.

H. *Amnicola retromargo* new species. Peters Creek, 3.5 mi. n.n.w. Green Cove Springs, Clay County, Florida.

Each scale equals 1 mm.

FIGURE 27

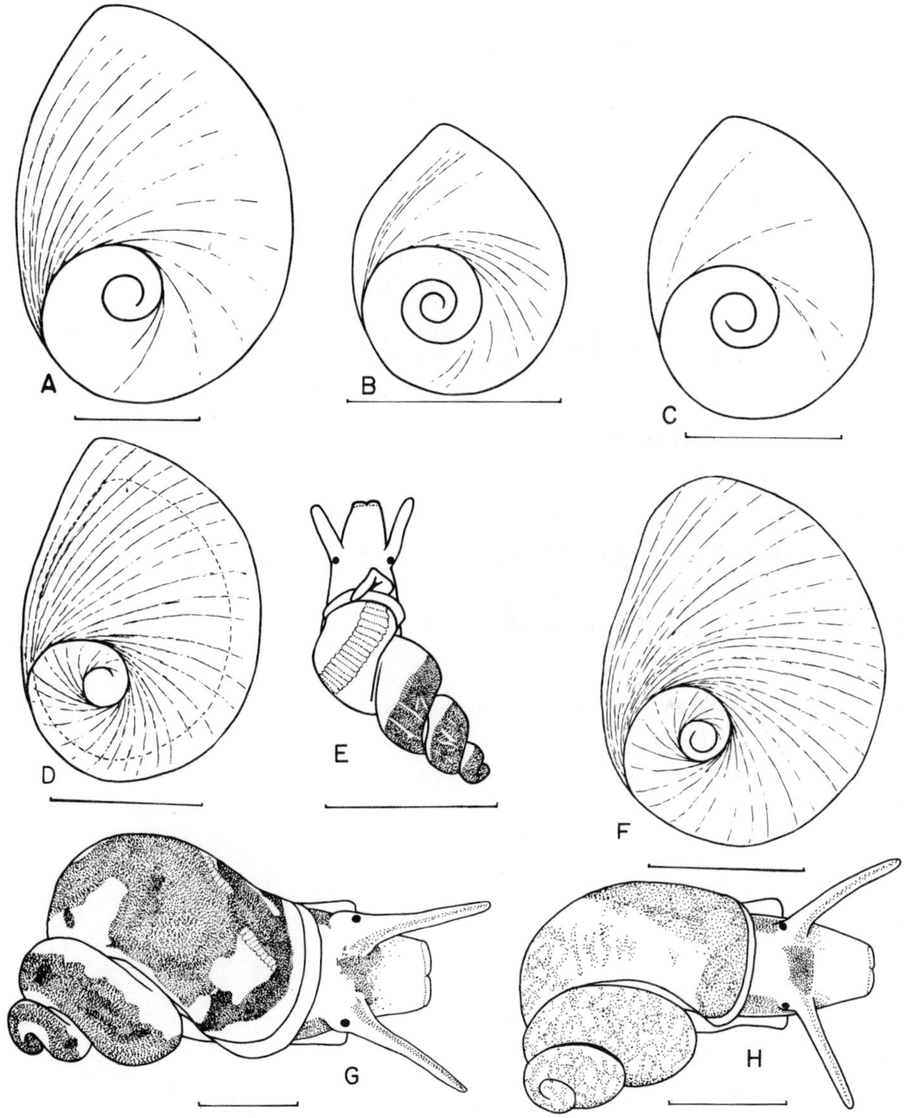

FIGURE 28

Opercula of various Florida hydrobiids

A. *Amnicola rhombostoma* new species.

B. *Amnicola d. dalli* (Pilsbry).

C. *Amnicola* (*Lyogyrus*) *retromargo* new species.

D. *Amnicola* (*Lyogyrus*) *granum* (Say).

E. *Amnicola dalli johnsoni* Pilsbry.

F. *Amnicola* (*Lyogyrus*) *pupoidea* (Gould).

G. *Onobops crassa* new species.

H. *Onobops jacksoni* (Bartsch).

Each scale equals 1 mm.

FIGURE 28

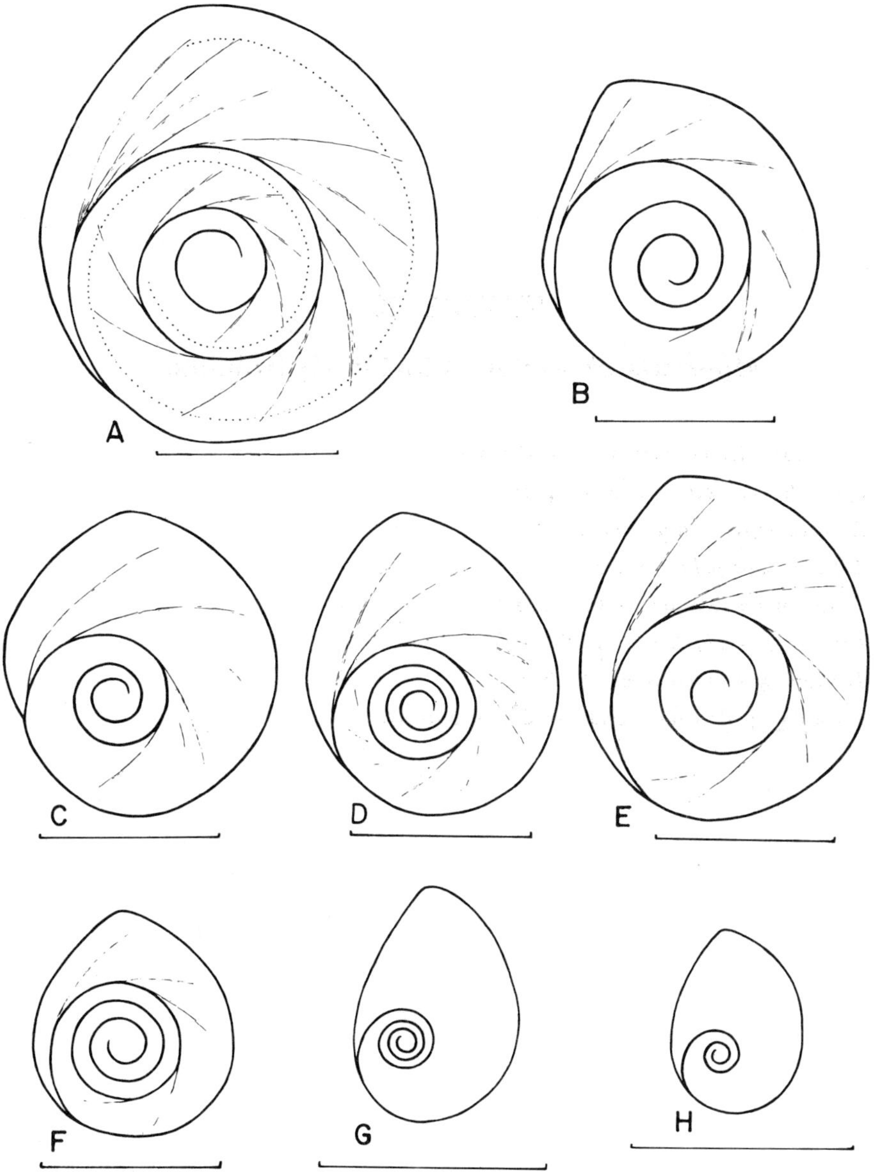

FIGURE 29

Opercula of various Florida Hydrobiidae

A. *Heleobops docima* new species. Female.

B. *Pyrgophorus platyrachis* new species.

C. *Littoridinops tenuipes* (Couper).

D. *Littoridinops palustris* new species.

E. *Littoridinops monroensis* (Frauenfeld).

F. *Hyalopyrgus aequicostatus* (Pilsbry).

G. *Hyalopyrgus brevissimus* (Pilsbry).

Each scale equals 1 mm; scale for C is also for D.

FIGURE 29

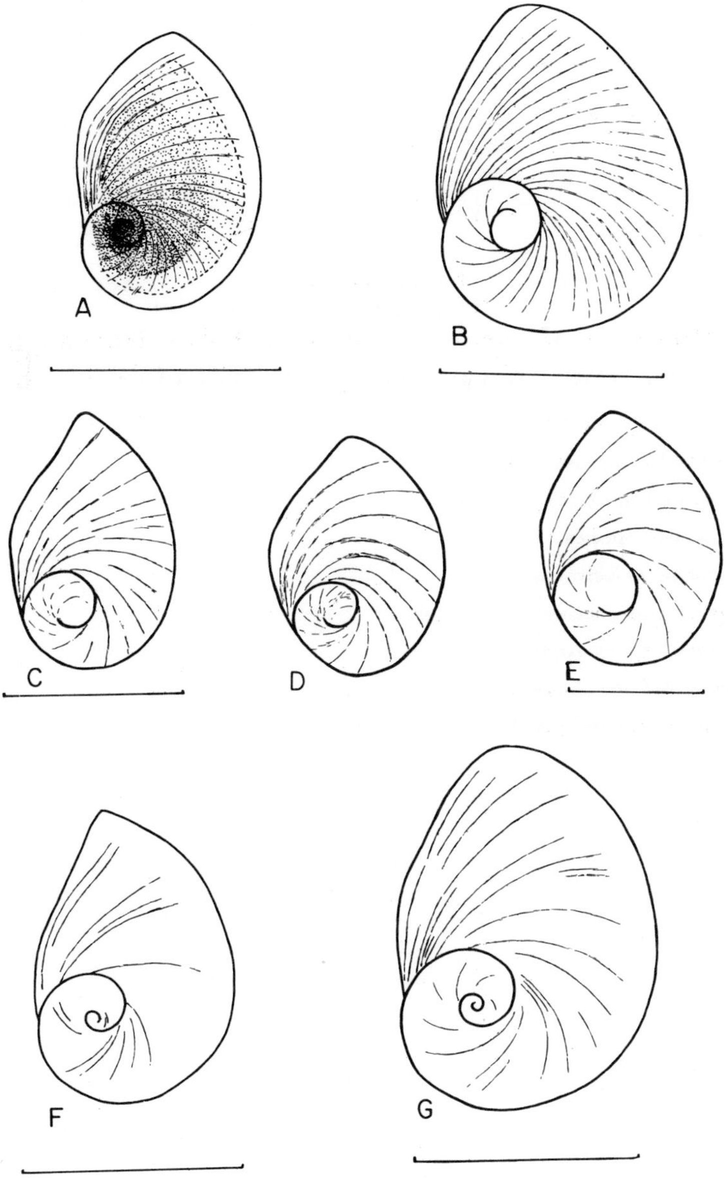

FIGURE 30

Opercula of various species of *Aphaostracon*

A. A. *rhadinus* new species.

B. A. *hypohyalina* new species.

C. A. *monas* (Pilsbry).

D. A. *pycnus* new species.

E. A. *theiocrenetus* new species.

F. A. *asthenes* new species.

G. A. *pachynotus* new species.

H. A. *xynoelictus* new species.

I. A. *chalarogyrus* new species.

Scale equals 1 mm.

FIGURE 30

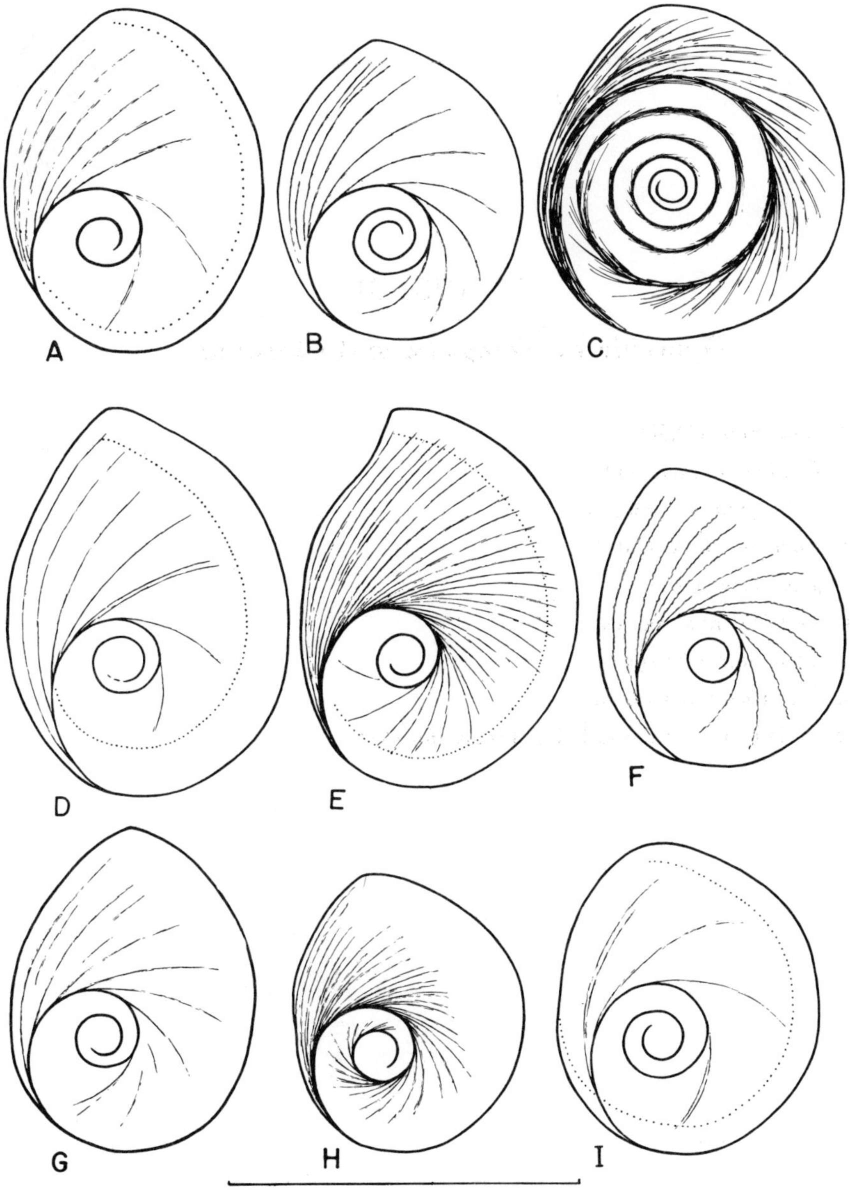

FIGURE 31

Opercula of *Notogillia* and *Cincinnatia*

A. *N. wetherbyi* (Dall).

B. *C. floridana* (Frauenfeld).

C. *C. helicogyra* new species.

D. *C. vanhyningi* (Vanatta).

E. *C. mica* new species.

F. *C. petrifons* new species.

G. *C. ponderosa* new species.

H. *C. wekiwae* new species.

Each scale equals 1 mm; scale for B is also for C-H.

FIGURE 31

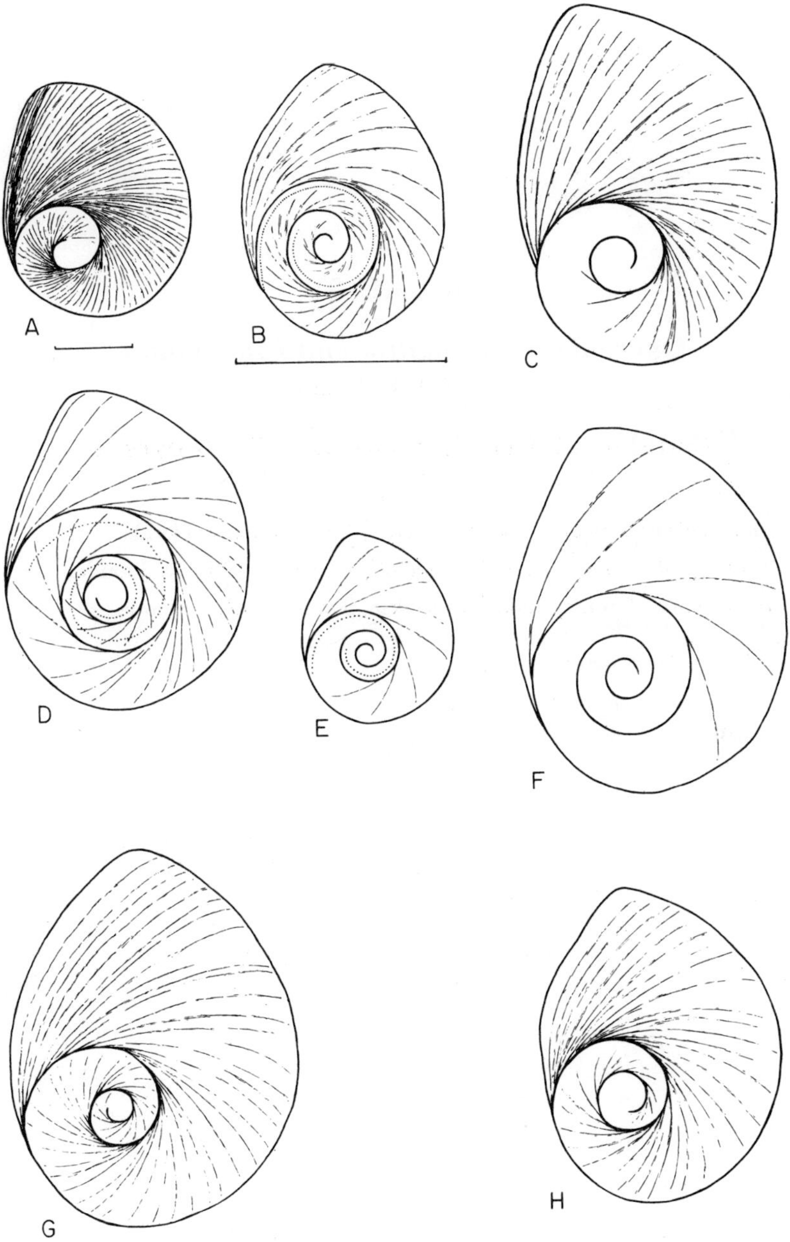

FIGURE 32

Radulae of *Pyrgophorus* and *Hyalopyrgus*

A. *P. platyrachis* new species. Canal, Ochopee, Collier County, Florida.

B. *H. aequicostatus* (Pilsbry). Lake Panasoffkee, Sumter County, Florida.

C. *H. brevissimus* (Pilsbry). Lake Panasoffkee, Sumter County, Florida.

Scale equals 10μ.

FIGURE 32

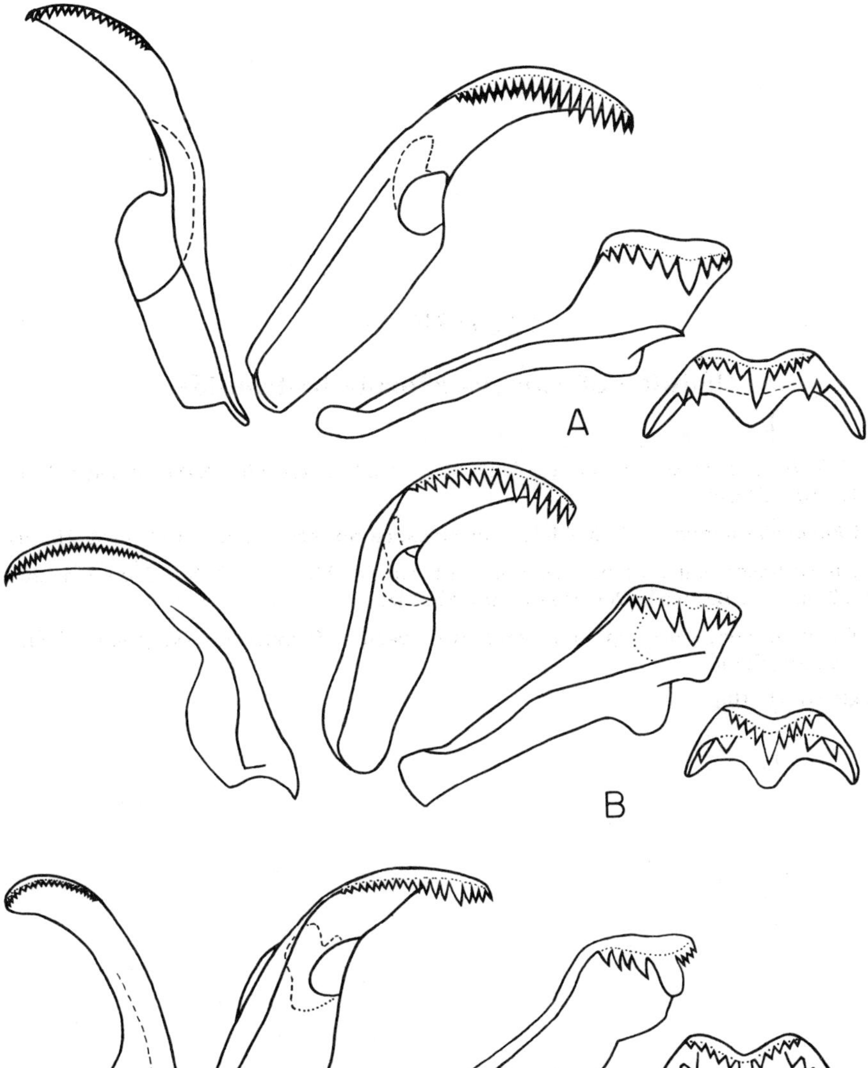

FIGURE 33

Radulae of various Florida hydrobiids

A. *Heleobops docima* new species. Nine-Mile Pond, Everglades National Park, Dade County, Florida.

B. *Cincinnatia floridana* (Frauenfeld). Ichetucknee Springs, Columbia County, Florida.

C. *Aphaostracon rhadinus* new species. Fish Creek, 3.3 mi. e. U.S. 17 on Fla. 209, 3.7 mi. s. of Bostwick, Putnam County, Florida.

D. *Onobops crassa* new species. Mangrove swamp 1.9 miles n. Everglades, Collier County, Florida.

Scale equals 10μ.

FIGURE 33

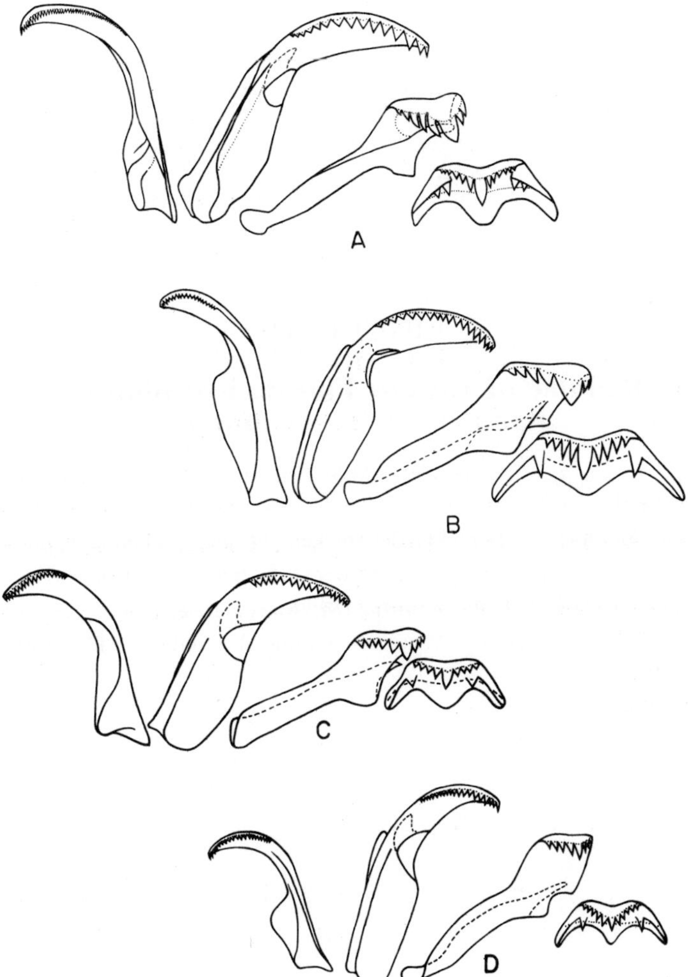

FIGURE 34

Radulae of *Littoridinops*

A. *L. palustris* new species. Marsh 3.2 mi. s. Yankeetown, Levy County, Florida.

B. *L. monroensis* (Frauenfeld). Marsh 0.6 mi. s.e. Keaton Beach, Taylor County, Florida.

C. *L. tenuipes* (Couper). Ditch entering St. Marys River, near U.S. 17, Nassau County, Florida.

Scale equals 10μ.

FIGURE 34

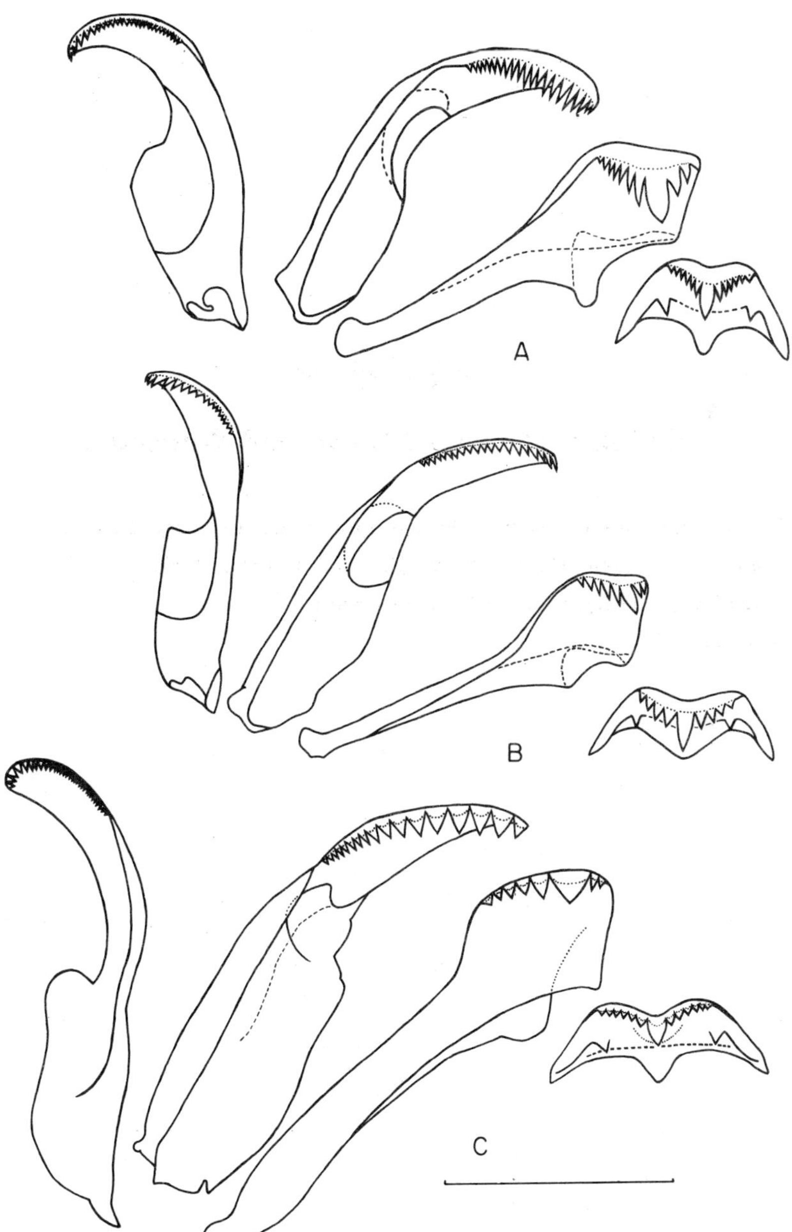

FIGURE 35

Radulae of *Spilochlamys* and *Notogillia*

A. S. *conica* new species. Natural Bridge Springs, Leon County, Florida.

B. S. *gravis* new species. Wekiva River, Seminole County, Florida.

C. *N. wetherbyi* (Dall). Lake Eustis, Lake County, Florida.

Scale equals 10μ.

FIGURE 35

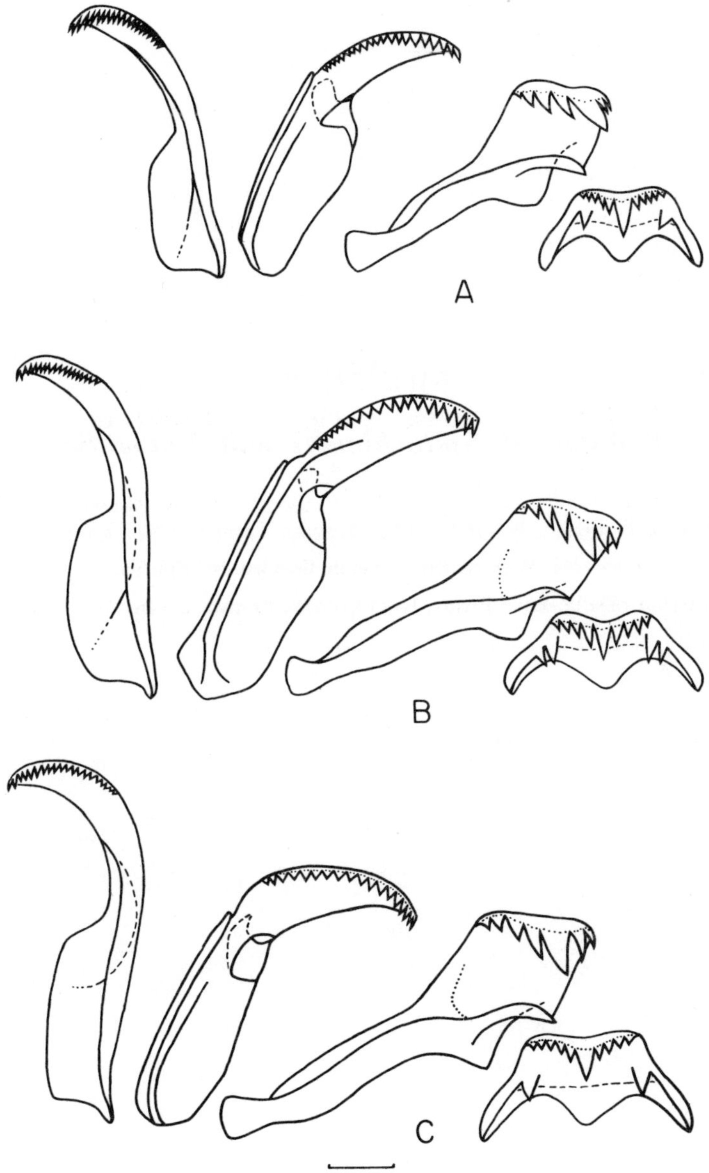

FIGURE 36

Radulae of three species of *Amnicola*

A. *A. rhombostoma* new species. Peters Creek, 3.5 mi. n.n.w. Green Cove Springs, Clay County, Florida.

B. *A. d. dalli* (Pilsbry). Wekiwa Springs, Orange County, Florida.

C. *A. retromargo* new species. Peters Creek, 3.5 mi. n.n.w. Green Cove Springs, Clay County, Florida.

Scale equals 50μ.

FIGURE 36

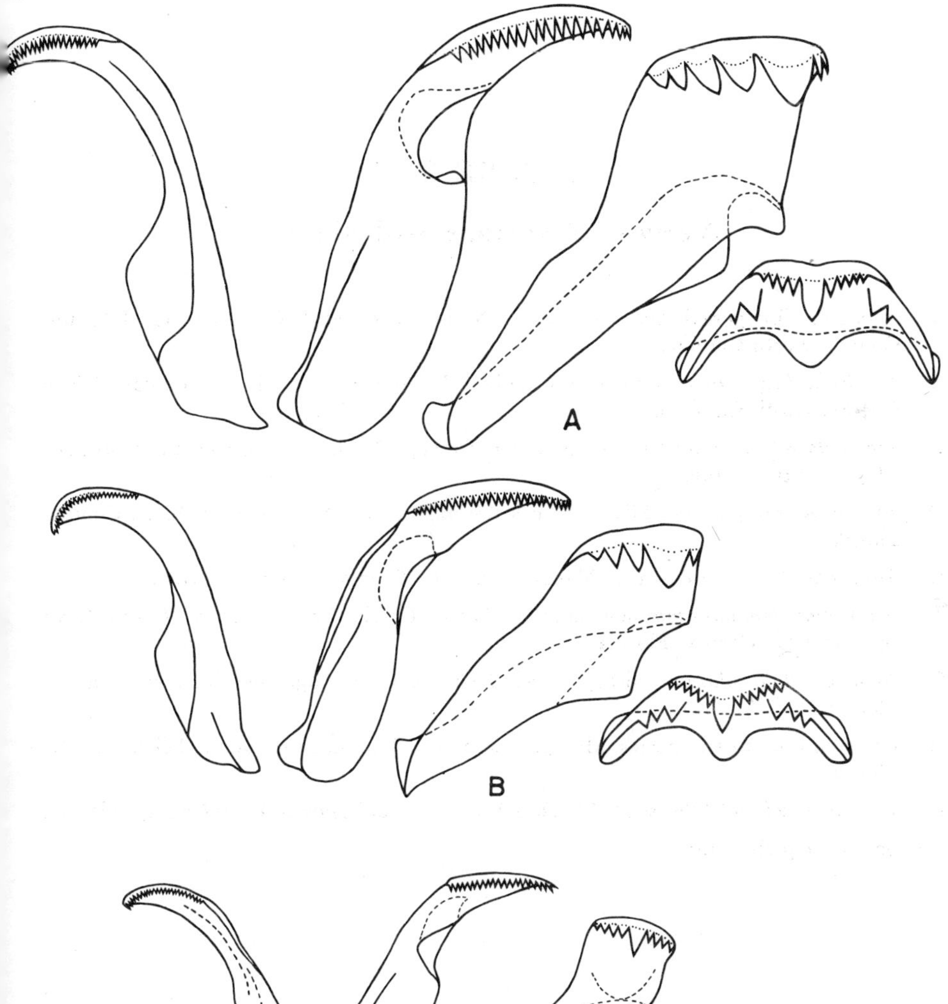

FIGURE 37

Verges of various hydrobiids

A. *Amnicola (Lyogyrus) granum* (Say). Neuse River at N.C. Hwy. 42, Johnston County, North Carolina.

B. *Amnicola (Lyogyrus) pupoidea* (Gould). Little River, 2 mi. w. Zebulon, Wake County, North Carolina.

C. *Amnicola retromargo* new species. Peters Creek, 3.5 mi. n.n.w. Green Cove Springs, Clay County, Florida.

D. *Amnicola dalli johnsoni* Pilsbry. Creek 2.9 mi. s.w. St. Augustine, St. Johns County, Florida.

E. *Amnicola d. dalli* (Pilsbry). Wekiwa Springs, Orange County, Florida.

F. *Amnicola rhombostoma* new species. Peters Creek, 3.5 mi. n.n.w. Green Cove Springs, Clay County, Florida.

G. *Amnicola dalli johnsoni* Pilsbry. Creek 2.9 mi. s.w. St. Augustine, St. Johns County, Florida.

H. *Onobops crassa* new species. Mangrove swamp 1.8 mi. n. Everglades, Collier County, Florida.

I. *Onobops jacksoni* (Bartsch). Marsh 3.2 mi. s.w. Yankeetown, Levy County, Florida.

Each scale equals 1 mm.

FIGURE 37

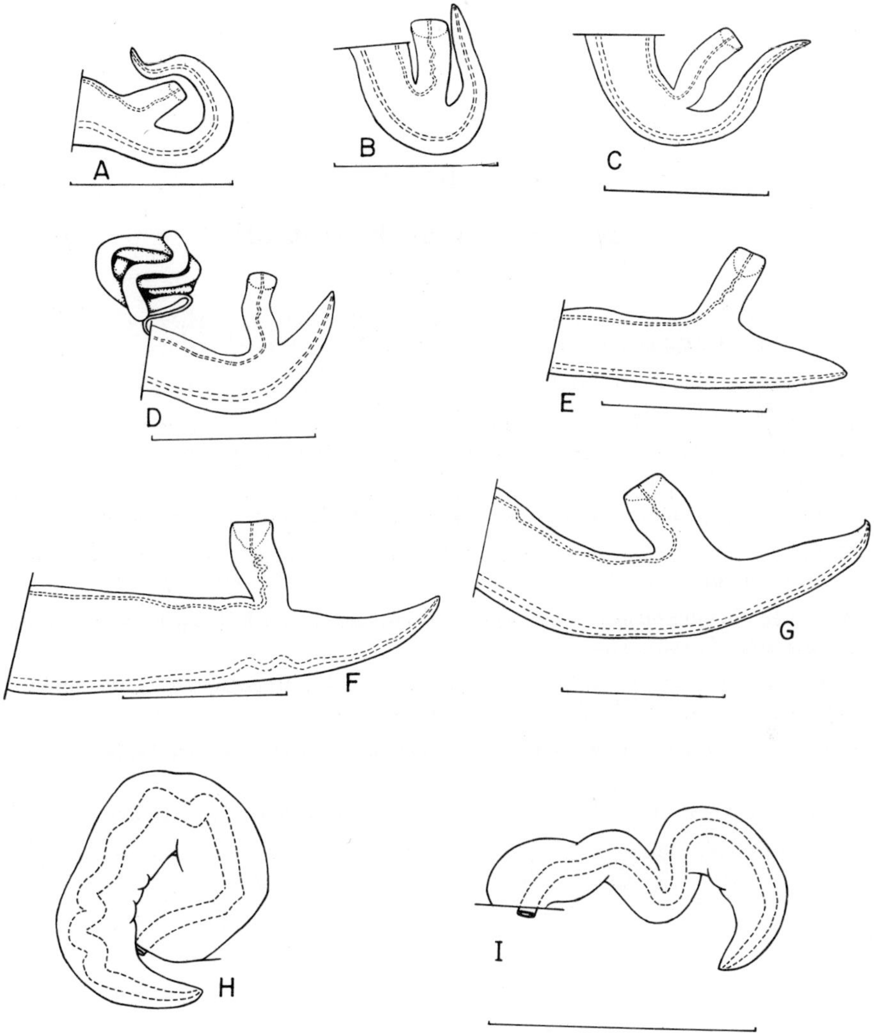

FIGURE 38

Verges of various Florida hydrobiids

A. *Heleobops docima* new species. Dorsal view of verge. Nine-Mile Pond, Everglades National Park, Dade County, Florida.

B. *Heleobops docima* new species. Dorsal view of verge. Marsh near the intersection of Fla. 29 and U.S. 41, Collier County, Florida.

C. *Heleobops docima* new species. Dorsal view of verge. Quarry, Big Pine Key, Monroe County, Florida.

D. *Littoridinops monroensis* (Frauenfeld). Cross section through a papillae from the right margin of the verge.

E. *Heleobops docima* new species. Longitudinal section through the right margin of the verge showing the structure of an apocrine gland.

Scale for A-C equals 1 mm; scale for D equals 50μ; scale for E equals 35μ.

FIGURE 38

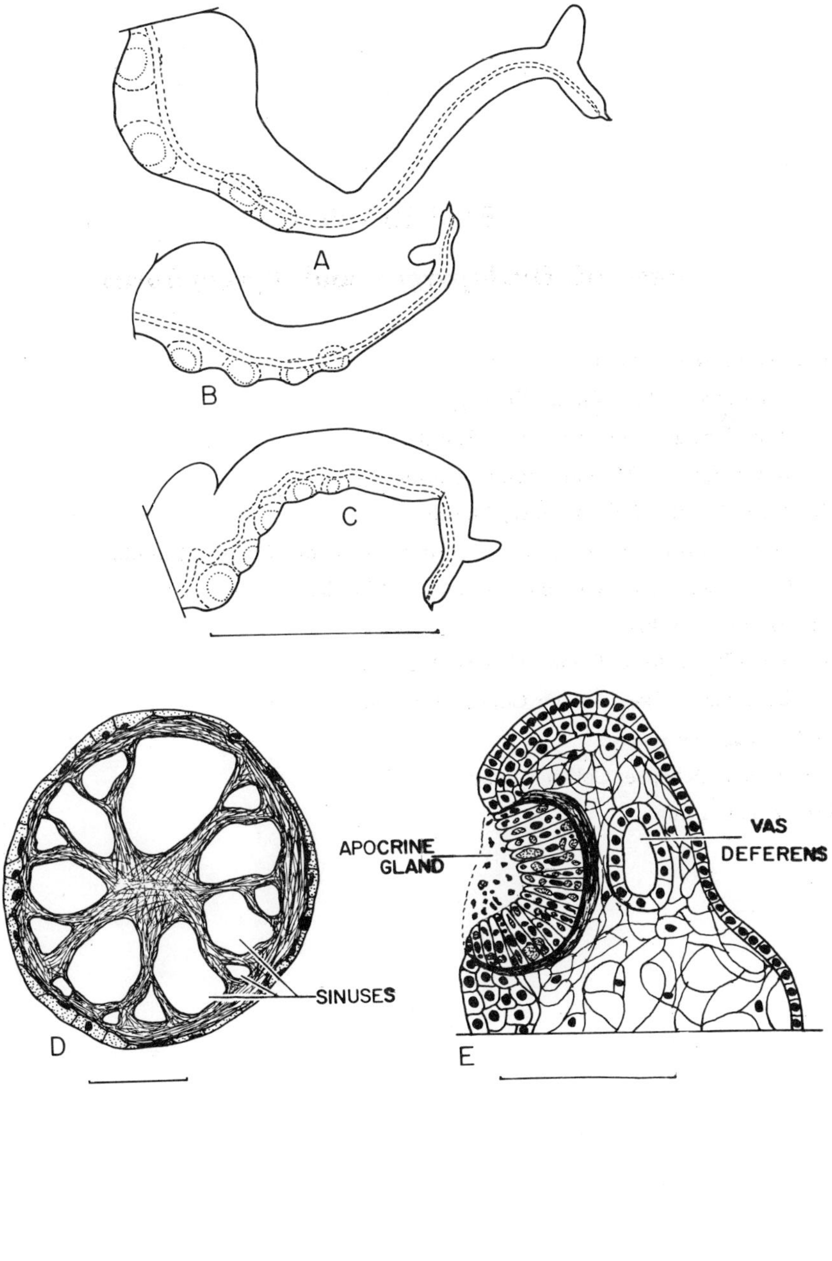

FIGURE 39

Verges of *Hyalopyrgus* and *Pyrgophorus*

H. aequicostatus (Pilsbry)

A. Lake Panasoffkee, Sumter County, Florida.

B. Blue Springs, Volusia County, Florida

C. Silver Springs, Marion County, Florida.

D. Lake Apopka, Lake County, Florida.

E. Sulfur Springs, near Sanlando Springs, Seminole County, Florida.

F. Lake Okeechobee, Okeechobee County, Florida.

H. brevissimus (Pilsbry)

G. Lake Panasoffkee, Sumter County, Florida.

H. Lake Panasoffkee, Sumter County, Florida.

P. platyrachis new species

I–K. Canal, Ochopee, Collier County, Florida.

Scale equals 1 mm.

FIGURE 39

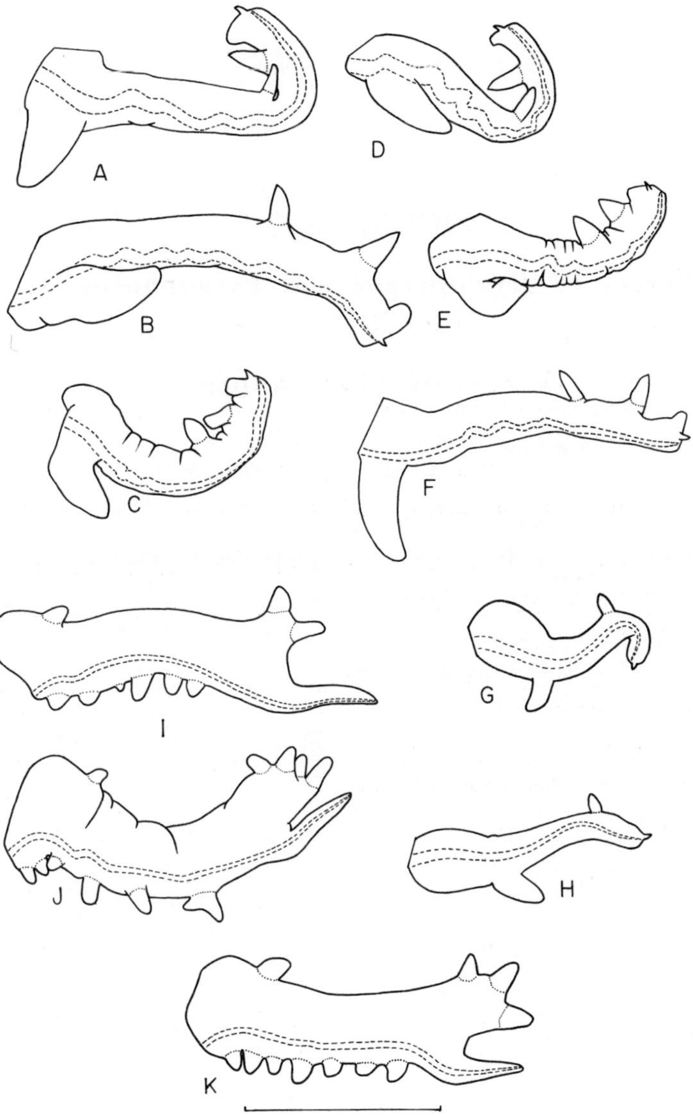

FIGURE 40

Verges of *Littoridinops*

A. *L. tenuipes* (Couper). Jonathan Dickinson State Park, Martin County, Florida.

B. *L. tenuipes* (Couper). Ditch entering St. Marys River, near U.S. 17, Nassau County, Florida.

C. *L. monroensis* (Frauenfeld). Pond on Merritt Island, Brevard County, Florida.

D. *L. monroensis* (Frauenfeld). Ditch 1 mi. n.e. West Lake, Everglades National Park, Dade County, Florida.

E–G. *L. palustris* new species. Marsh 3.2 mi. s.w. Yankeetown, Levy County, Florida.

Scale equals 1 mm.

FIGURE 40

FIGURE 41

Verges of *Aphaostracon*

A,B. *A. rhadinus* new species. Fish Creek, 3.7 mi. s. Bostwick, Putnam County, Florida.

C. *A. monas* (Pilsbry). Wekiwa Springs, Orange County; Florida.

D,E. *A. hypohyalina* new species. Mill Spring, Ichetucknee Springs, Columbia County, Florida.

F. *A. pycnus* new species. Alexander Spring Run, Lake County, Florida.

G,H. *A. theiocrenetus* new species. Clifton Springs, Seminole County, Florida.

I. *A. asthenes* new species. Blue Springs, 3 mi. w. Orange, Volusia County, Florida.

J. *A. pachynotus* new species. Canal 2.5 mi. north Dade County Line, U.S. Hwy. 27, Broward County, Florida.

K. *A. xynoelictus* new species. Fenney Springs, 2 mi. e. Coleman, Sumter County, Florida.

L,M. *A. chalarogyrus* new species. Magnesia Springs, 3.7 mi. w. Hawthorne, Alachua County, Florida.

Scale equals 1 mm.

FIGURE 41

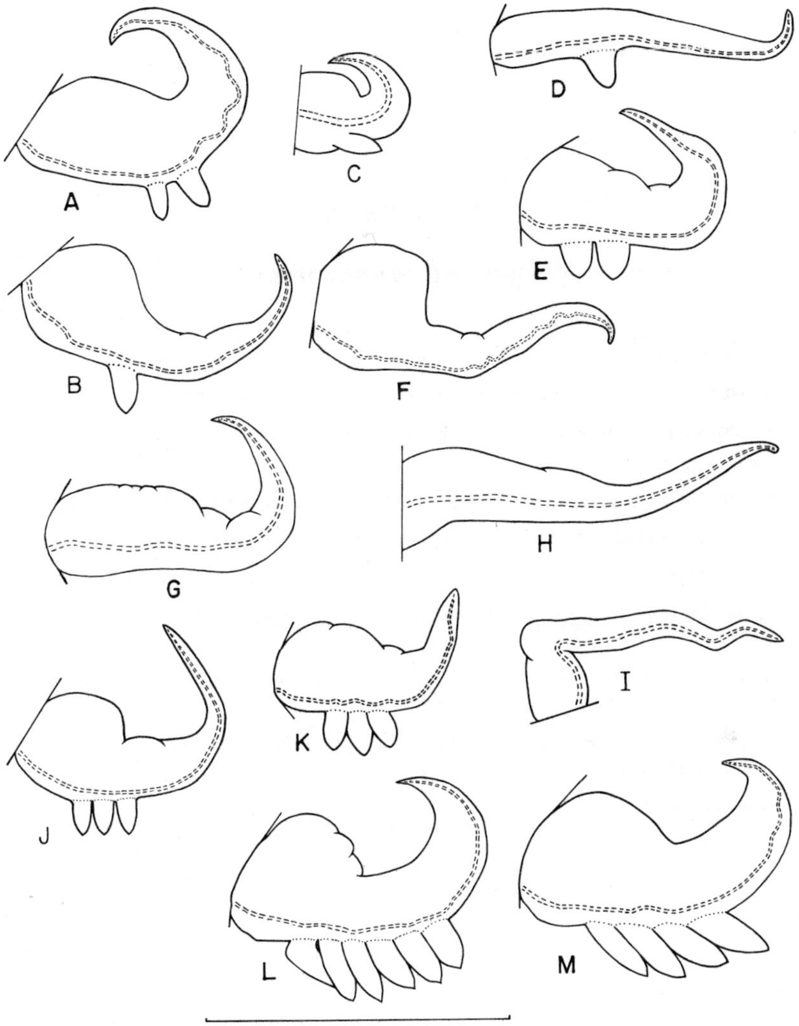

FIGURE 42

Verges of *Spilochlamys* and *Notogillia*

S. *conica* new species

A. Dorsal surface. Natural Bridge Springs, Leon County, Florida.

B. Ventral surface of A.

S. *gravis* new species

C. Dorsal surface. Wekiva River, Seminole County, Florida.

D. Ventral surface of C.

N. wetherbyi (Dall)

E. Dorsal surface. Manatee Springs, Levy County, Florida.

F. Ventral surface of E.

G. Lateral view of E from right margin.

H. Dorsal surface. Lake Eustis, Lake County, Florida.

I. Ventral surface of H.

Each scale equals 1 mm.

FIGURE 42

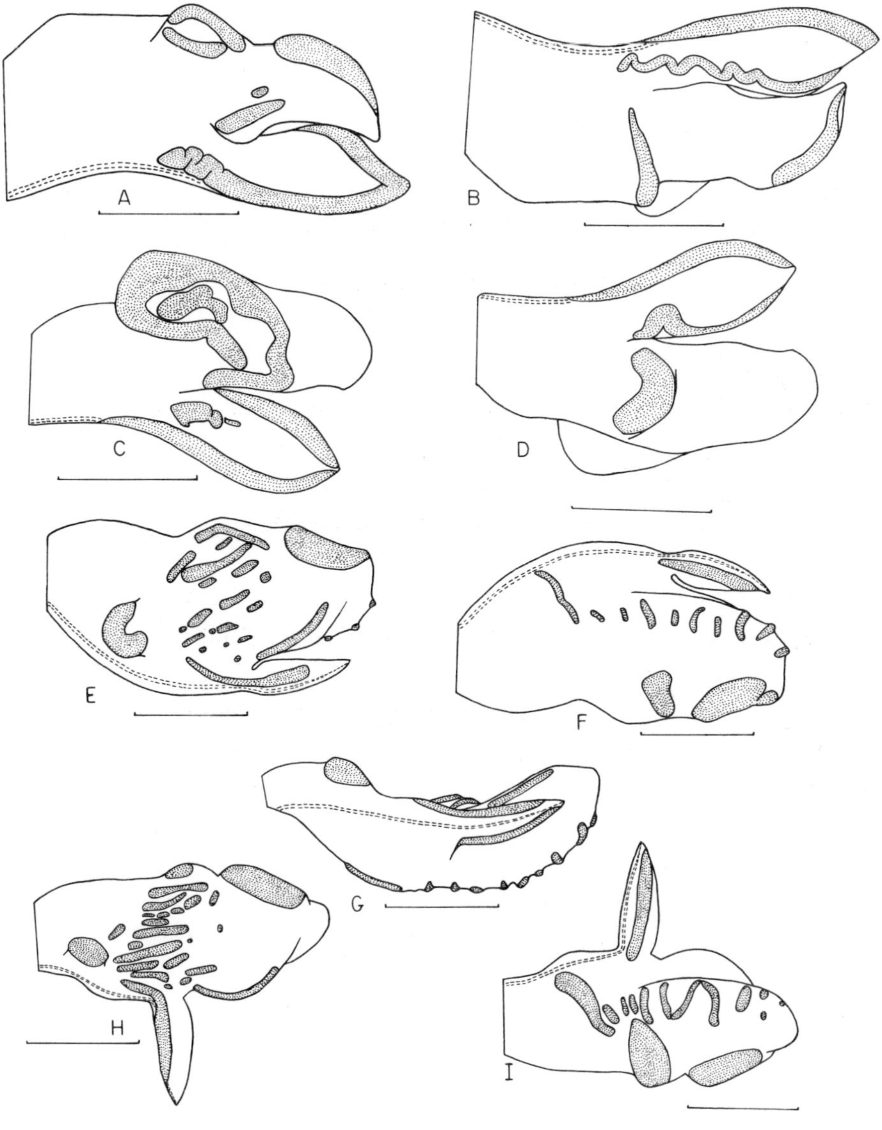

FIGURE 43

Verges of *Cincinnatia*

C. floridana (Frauenfeld)

A. Dorsal surface. Ponte Vedra Beach, St. Johns County, Florida.

B. Dorsal surface. Welaka Springs, Putnam County, Florida.

C. Dorsal surface. Welaka Springs, Putnam County, Florida.

D. Dorsal surface. Withlacoochee River, Citrus County, Florida.

E. Ventral view of D.

C. mica new species

F. Dorsal surface. Ichetucknee Springs, Suwannee County, Florida.

G. Ventral surface of F.

H. Dorsal surface. Ichetucknee Springs, Suwannee County, Florida.

I. Ventral surface of H.

Scale equals 1 mm.

FIGURE 43

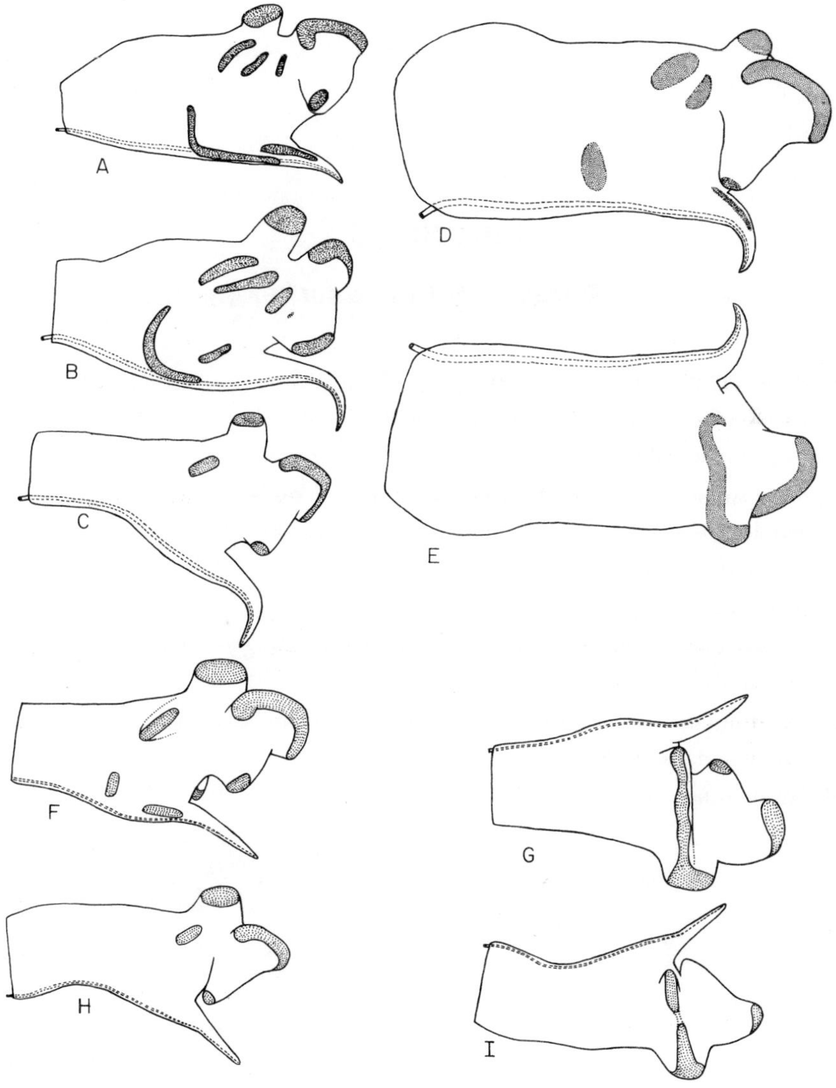

FIGURE 44

Verges of three species of *Cincinnatia*

C. monroensis (Dall). Spring run 1 mi. n. Enterprise, Volusia County, Florida.

A. Outer surface.

B. Inner surface of A.

C. parva new species. Blue Springs, 3 mi. w. Orange, Volusia County, Florida.

C. Inner surface.

D. Outer surface of C.

E. Outer surface.

C. fraterna new species. Creek 4 mi. s. Tocoi, St. Johns County, Florida.

F. Outer surface.

G. Outer surface.

H. Inner surface of G.

Each scale equals 1 mm.

FIGURE 44

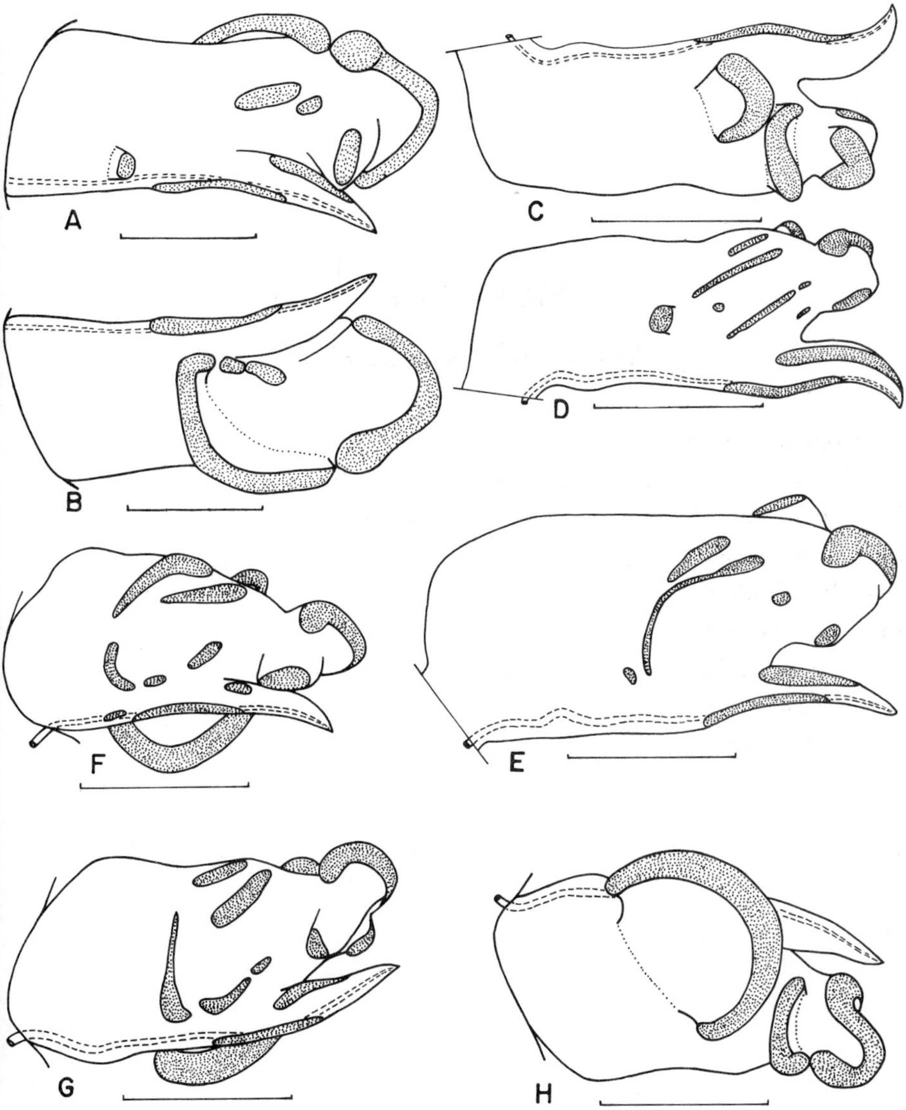

FIGURE 45

Verges of *Cincinnatia*

C. helicogyra new species

A. Dorsal surface. Crystal River, Citrus County, Florida.

B. Dorsal surface. Crystal River, Citrus County, Florida.

C. Ventral surface of A.

D. Ventral surface of C.

C. vanhyningi (Vanatta)

E–G. Dorsal surfaces. Seminole Springs, Lake County, Florida.

H. Ventral surface of G.

Scale equals 1 mm.

FIGURE 45

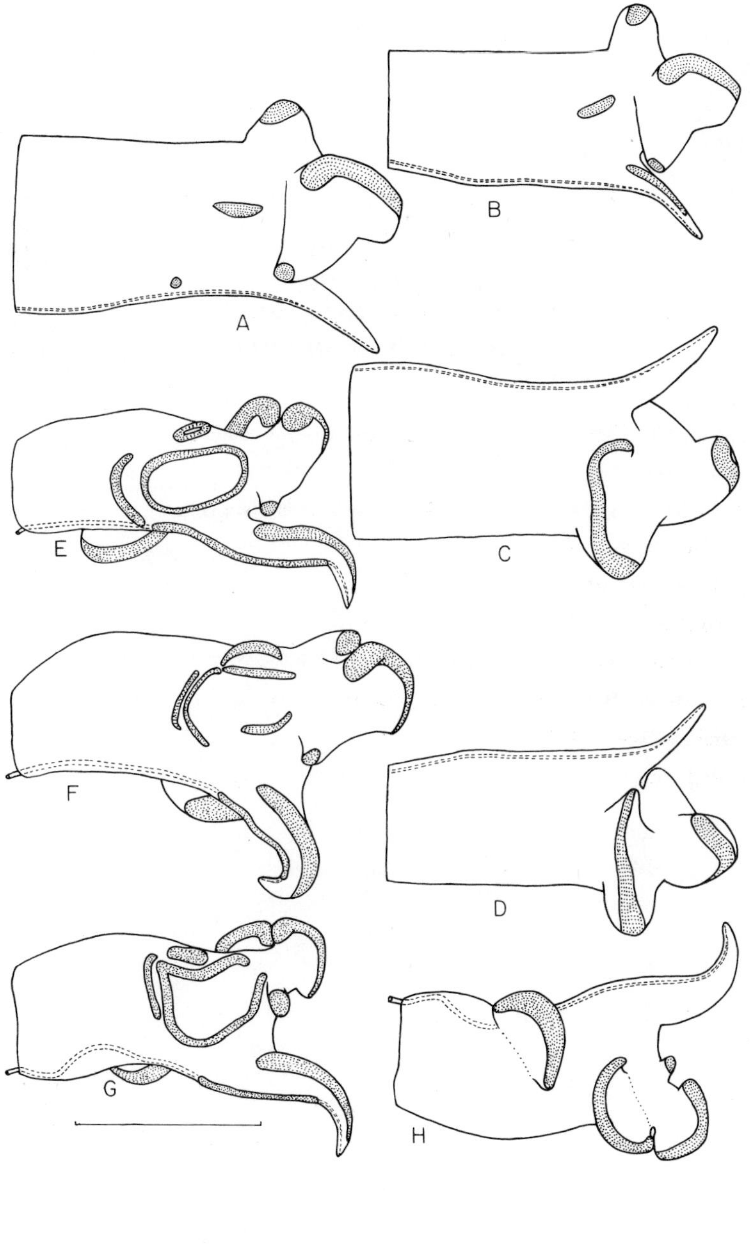

FIGURE 46

Verges of *Cincinnatia*

C. ponderosa new species

A. Dorsal surface. Sanlando Springs, Seminole County, Florida.

B. Ventral surface of A.

C. Dorsal surface. Sanlando Springs, Seminole County, Florida.

C. petrifons new species

D. Dorsal surface. Rock Springs, Orange County, Florida.

E. Ventral surface of G.

F. Dorsal surface. Rock Springs, Orange County, Florida.

G. Dorsal surface. Rock Springs, Orange County, Florida.

H. Ventral surface of D.

Scale equals 1 mm.

FIGURE 46

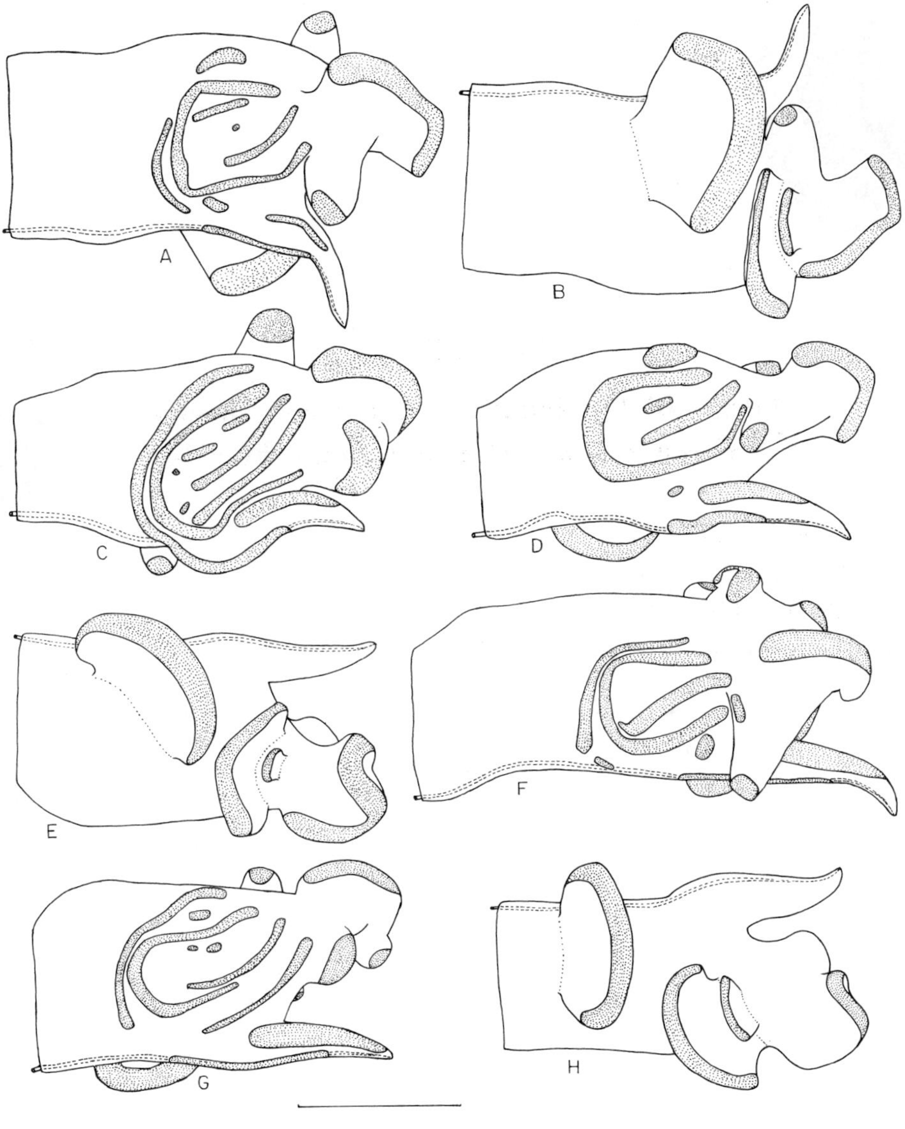

FIGURE 47

Verges of *Cincinnatia wekiwae* new species
Wekiwa Springs, Orange County, Florida

A. Dorsal surface.

B. Ventral surface of A.

C. Dorsal surface.

D. Ventral surface of C.

Scale equals 1 mm.

FIGURE 47

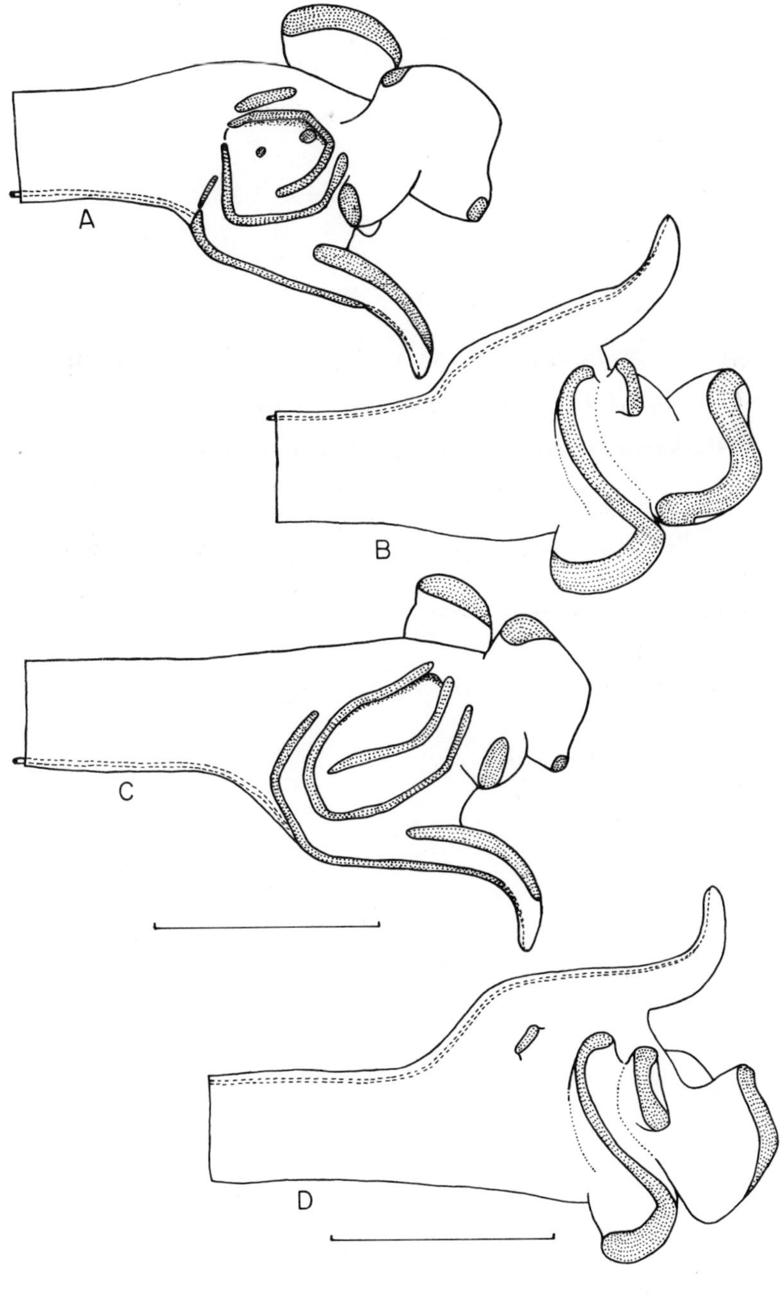

FIGURE 48

Shells of *Heleobops* new genus and *Littoridinops*

H. docima new species

A,B,E. Nine-Mile Pond, Everglades National Park, Dade County, Florida. Paratypes (*UMMZ* 216491).

C. Marsh 5.2 mi. w.n.w. intersection U.S. 41 and Florida Hwy. 29, Collier County, Florida.

F. Marsh 1.5 mi. n. Everglades, Collier County, Florida.

L. monroensis (Frauenfeld)

D. West Lake, Everglades National Park, Dade County, Florida.

Scale equals 1 mm.

FIGURE 48

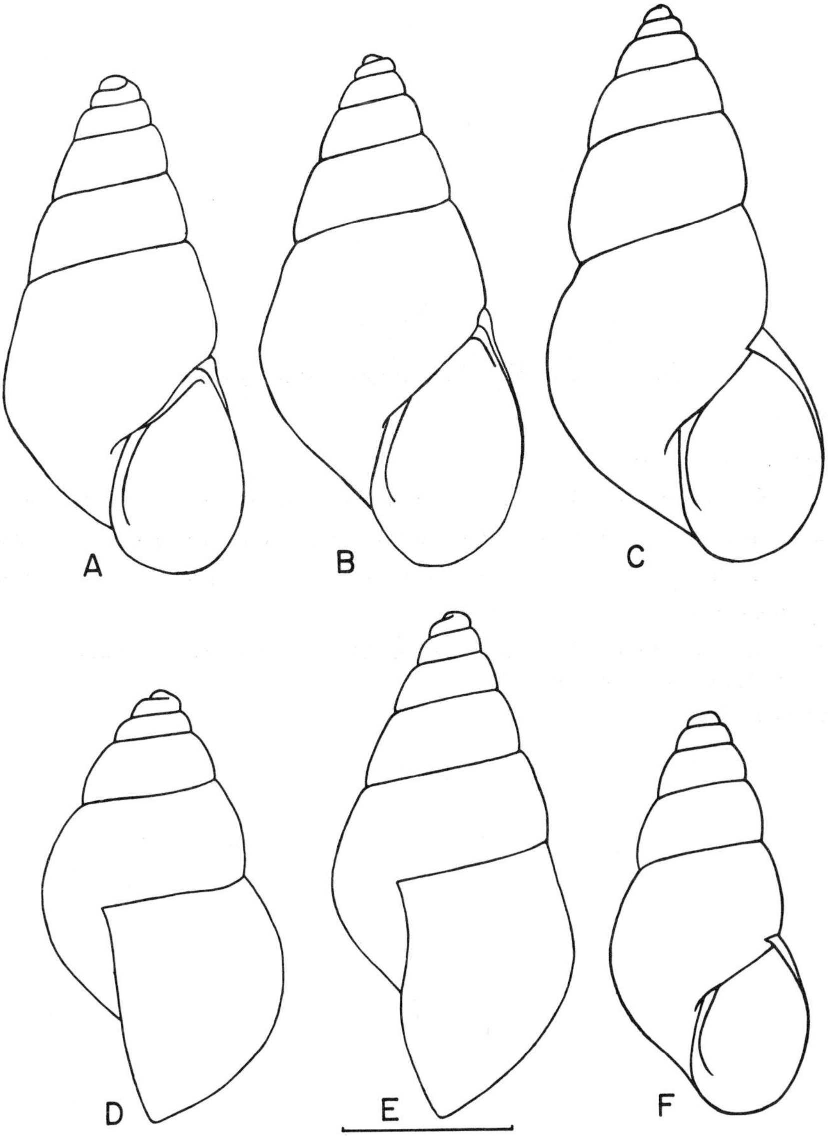

FIGURE 49

Shells of *Hyalopyrgus* new genus

H. aequicostatus (Pilsbry)

A. Lake Panasoffkee, 3.7 mi. n.n.w. Panasoffkee, Sumter County, Florida. Female.

B,D. Creek 7.4 mi. s.e. Geneva, Seminole County, Florida. B. Female. D. Male.

C,E. Blue Springs, 3 mi. e. Orange, Volusia County, Florida. C. Female. E. Male.

H. brevissimus (Pilsbry)

F–I. Lake Panasoffkee, 3.7 mi. n.n.w. Panasoffkee, Sumter County, Florida. F. Male. G–I. Females.

Scale equals 1 mm.

FIGURE 49

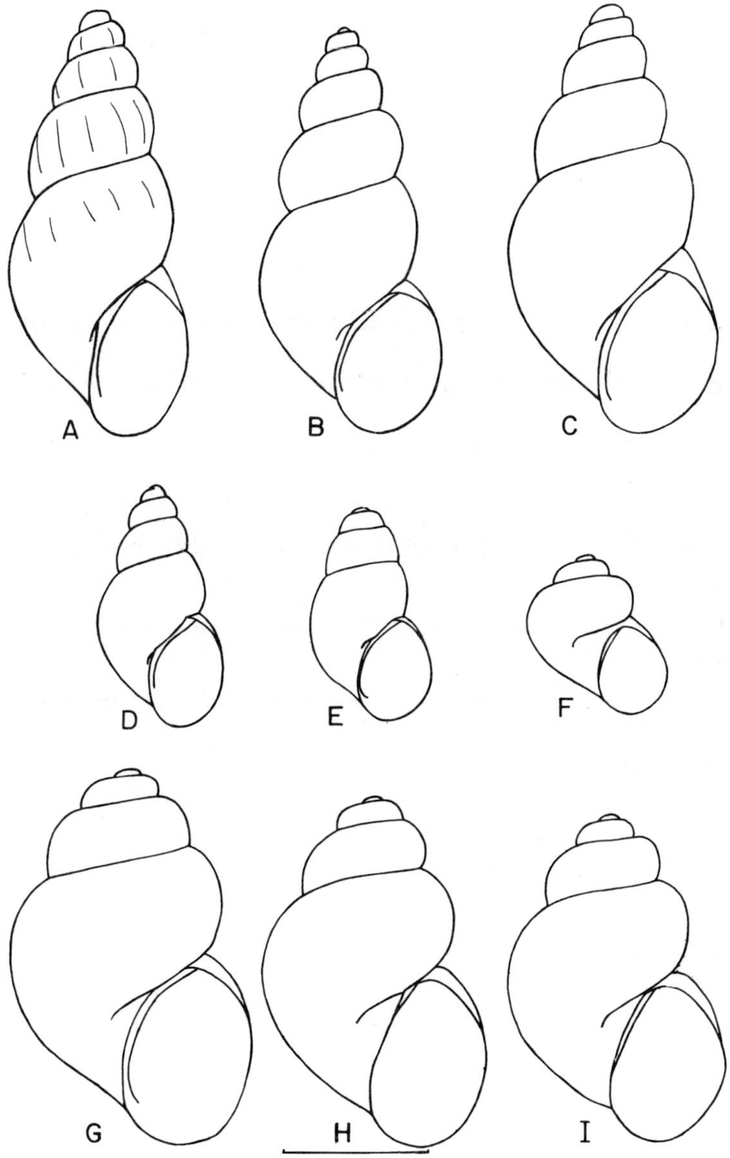

FIGURE 50

Shells of *Littoridinops* and *Onobops* new genus

L. tenuipes (Couper)

A–C. Kitchen Creek, Jonathan Dickinson State Park, Martin County, Florida.

L. palustris new species

D–F. Brackish marsh 3.2 mi. s.w. Yankeetown, Levy County, Florida. D,E. Paratypes (*UMMZ* 216493). F. Type (*UMMZ* 216492).

O. jacksoni (Bartsch)

G,H. Brackish marsh 3.2 mi. s.w. Yankeetown, Levy County, Florida.

O. crassa new species

I,J. Marsh 1.9 mi. n. Everglades, Monroe County, Florida. Paratypes (*UMMZ* 216538).

Each scale equals 1 mm.

FIGURE 50

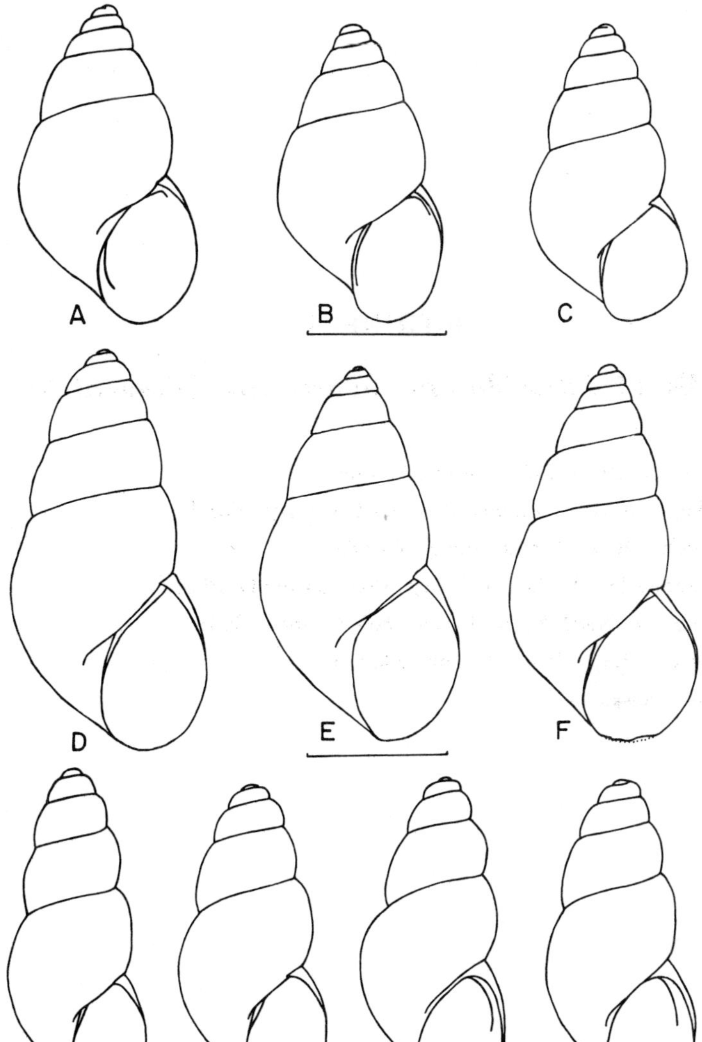

FIGURE 51

Shells of *Littoridinops monroensis* (Frauenfeld)

A. St. Johns River, Astor, Lake County, Florida.

B. West Lake, Everglades National Park, Dade County, Florida.

C. Withlacoochee River, Citrus County, Florida.

D. Marsh, Section 18, Middle Torch Key, Monroe County, Florida.

E. Quarry, 3 mi. w. North Miami Beach, Dade County, Florida.

F. Pond, Big Pine Key, Monroe County, Florida.

Scale equals 1 mm.

FIGURE 51

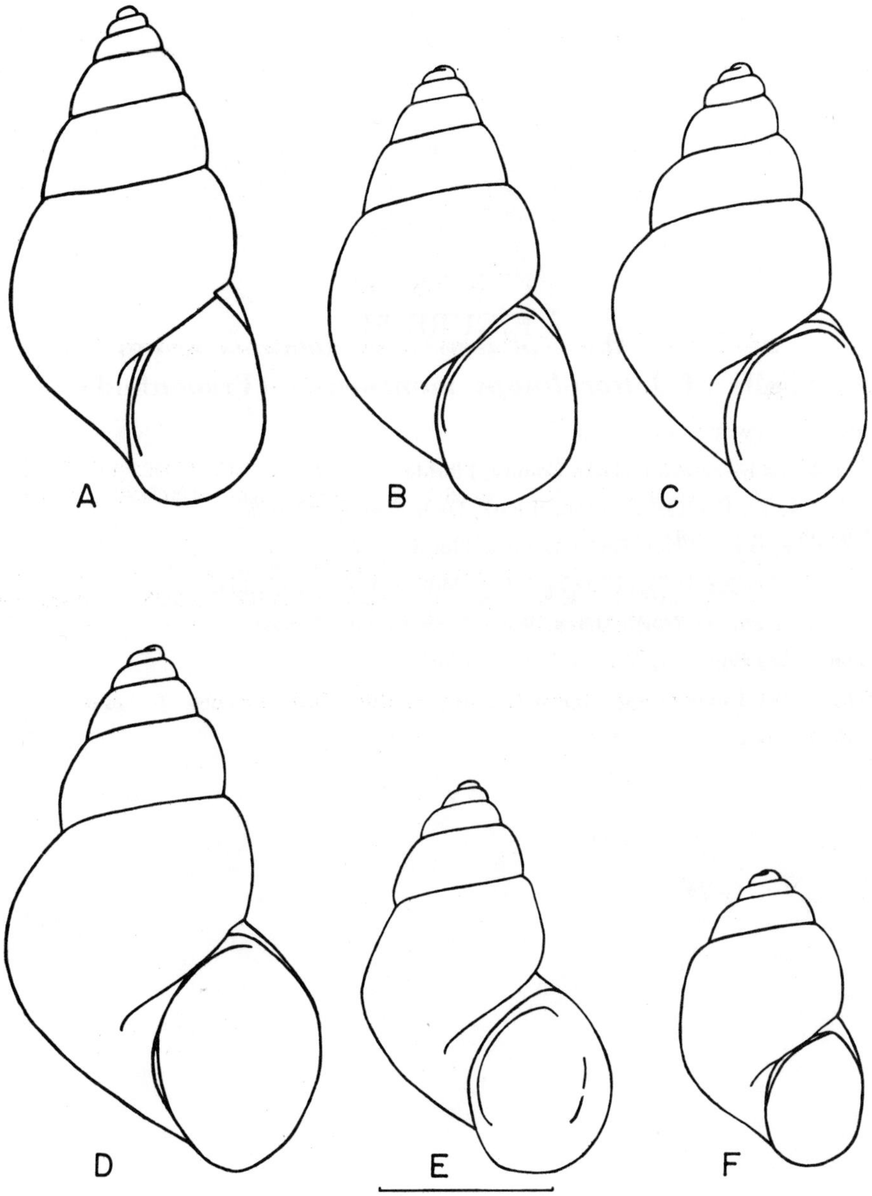

FIGURE 52

Shells of the *Aphaostracon rhadinus* group

A. *rhadinus* new species

A–C,K. Fish Creek, Putnam County, Florida. A. Type (*UMMZ* 216525). B,C. Paratypes, females (*UMMZ* 216526). K. Paratype, male (*UMMZ* 216526).

A. *hypohyalina* new species

D–F,I. Mill Spring, Ichetucknee Springs, Columbia County, Florida. D. Type (*UMMZ* 216527). E,F. Paratypes, females (*UMMZ* 216528). I. Paratype, male (*UMMZ* 216528).

A. *monas* (Pilsbry)

G,H,J. Wekiwa Springs, Orange County, Florida. G,H. Females. J. Male.

Scale equals 1 mm.

FIGURE 52

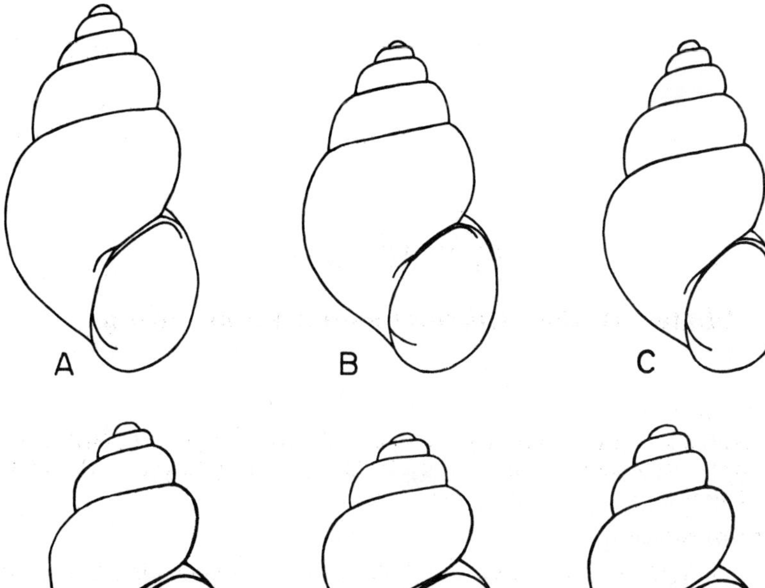

FIGURE 53

Shells of the *Aphaostracon pycnus* group

A. *pycnus* new species

A–C,J. Alexander Springs Run, Lake County, Florida. A. Type (*UMMZ* 216512). B,C. Paratypes, females (*UMMZ* 216513). J. Paratype, male (*UMMZ* 216514).

A. *asthenes* new species

D–F. Blue Springs, 3 mi. w. Orange, Volusia County, Florida. D. Type (*UMMZ* 216519). E, F. Paratypes, females (*UMMZ* 216520).

A. *theiocrenetus* new species

G–I,K. Clifton Springs, Seminole County, Florida. G. Type (*UMMZ* 216510). H,I. Paratypes, females (*UMMZ* 216511). K. Paratype, male (*UMMZ* 216511).

Scale equals 1 mm.

FIGURE 53

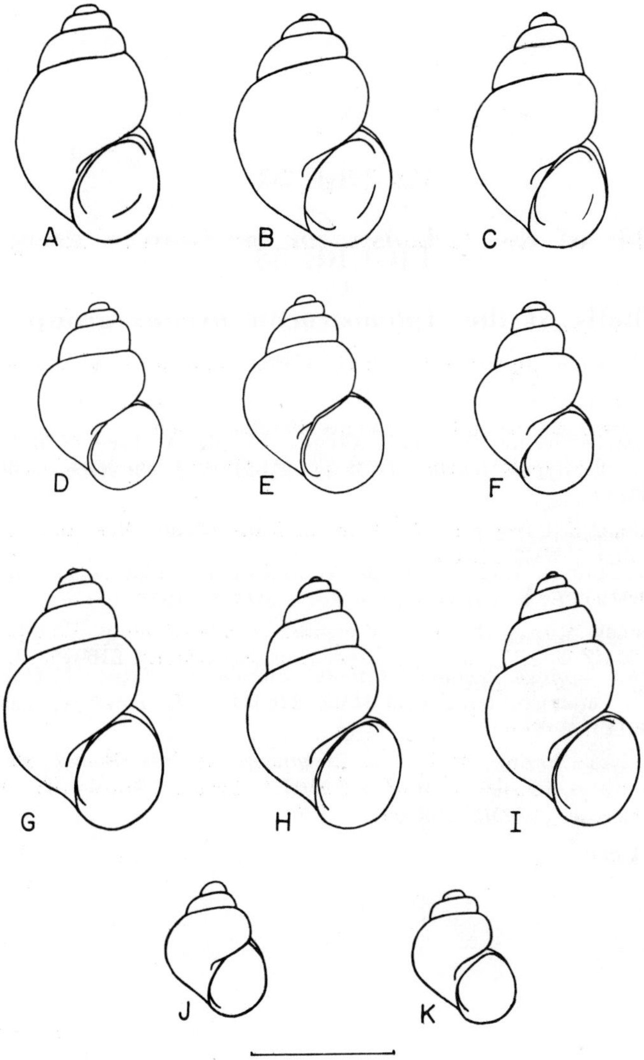

FIGURE 54

Shells of the *Aphaostracon pachynotus* group

A. *pachynotus* new species

A. Canal 7.4 mi. w. Jupiter, Palm Beach County, Florida. Type, female (*UMMZ* 216521).

B. Blue Cypress Lake, Indian River County, Florida. Female.

C. Canal 7.4 mi. w. Jupiter, Palm Beach County, Florida. Paratype, female (*UMMZ* 216522).

M. Canal along U.S. Highway 27, 6 mi. n. Dade County line, Broward County, Florida. Male (*UMMZ* 216524).

A. *xynoelictus* new species

D–F,L. Fenney Springs, 2 mi. e. Coleman, Sumter County, Florida. D. Type (*UMMZ* 216517). E,F. Paratypes, females (*UMMZ* 216518). L. Paratype, male (*UMMZ* 216518).

A. *chalarogyrus* new species

G–K. Magnesia Springs, 3.7 mi. w. Hawthorne, Alachua County, Florida. G–I. Paratypes, females (*UMMZ* 216516). J. Type (*UMMZ* 216515). K. Paratype, male (*UMMZ* 216516).

Scale equals 1 mm.

FIGURE 54

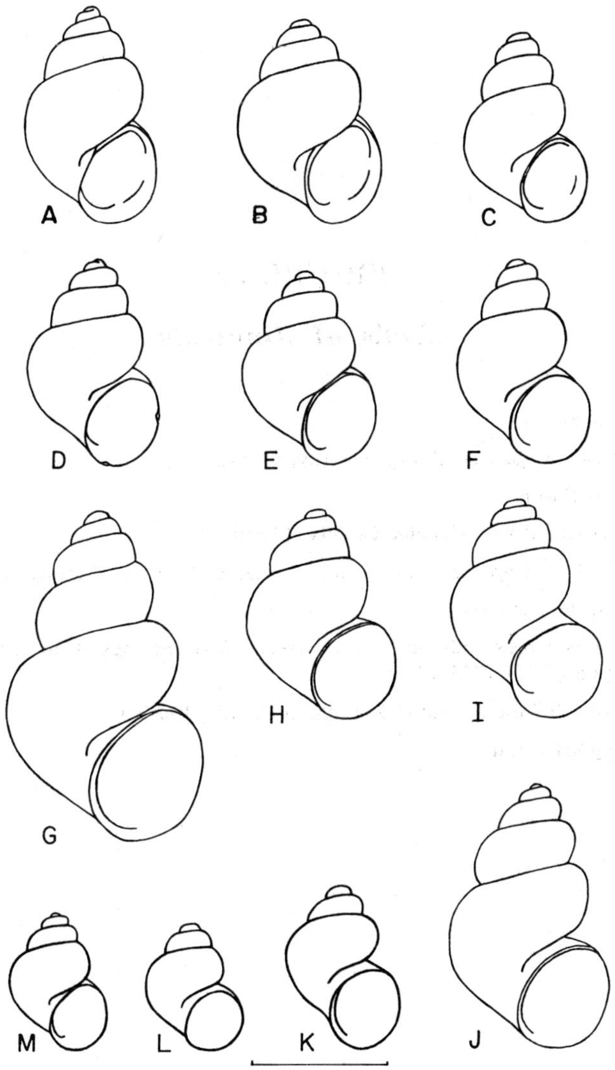

FIGURE 55

Shells of *Amnicola*

A. d. dalli (Pilsbry)

A–C. Wekiwa Springs, Orange County, Florida.

A. dalli johnsoni Pilsbry

D. Myacca River, Manatee County, Florida.

E,F. Heilbronn Springs, 5.6 mi. n.n.w. Starke, Bradford County, Florida.

A. rhombostoma new species

G. Peters Creek, 3.5 mi. n.n.w. Green Cove Springs, Clay County, Florida. Type (*UMMZ* 216505).

H. Creek 3.7 mi. s. Bostwick, Putnam County, Florida.

Each scale equals 1 mm.

FIGURE 55

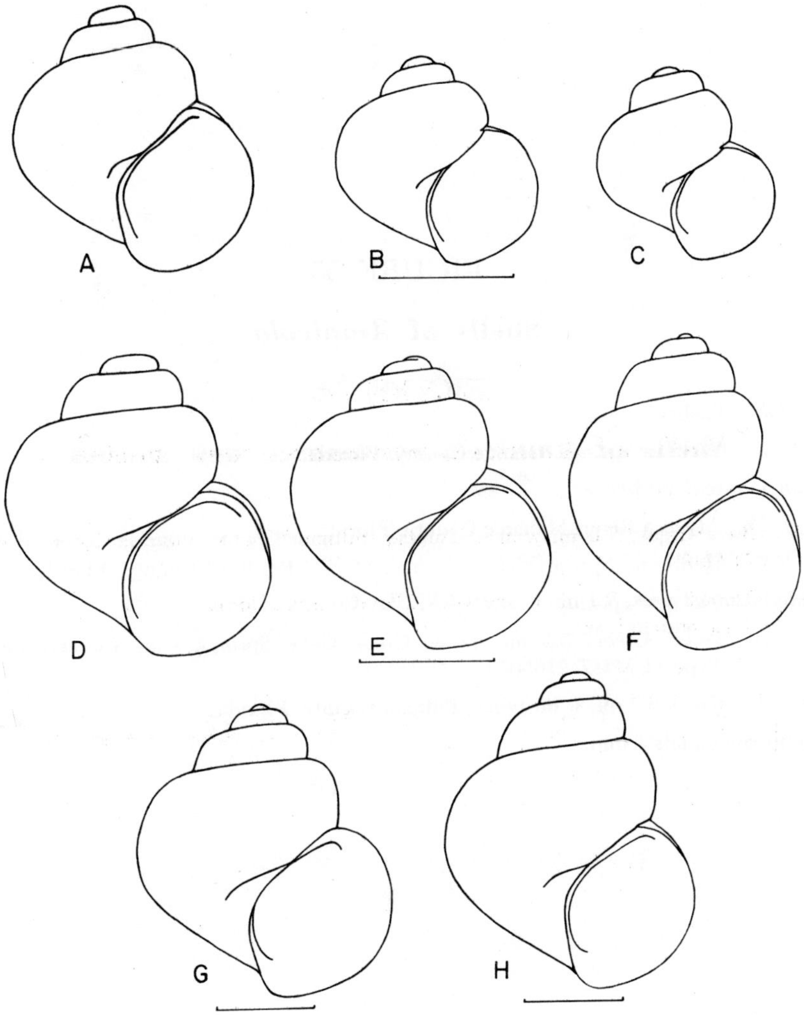

FIGURE 56

Shells of *Amnicola retromargo* new species

A,D–F. Rice Creek, 7.3 mi. w.n.w. Palatka, Putnam County, Florida. Type (*UMMZ* 216507).

B–C. Clark Creek, 9.1 mi. n. Bostwick, Clay County, Florida.
Scale equals 1 mm.

FIGURE 56

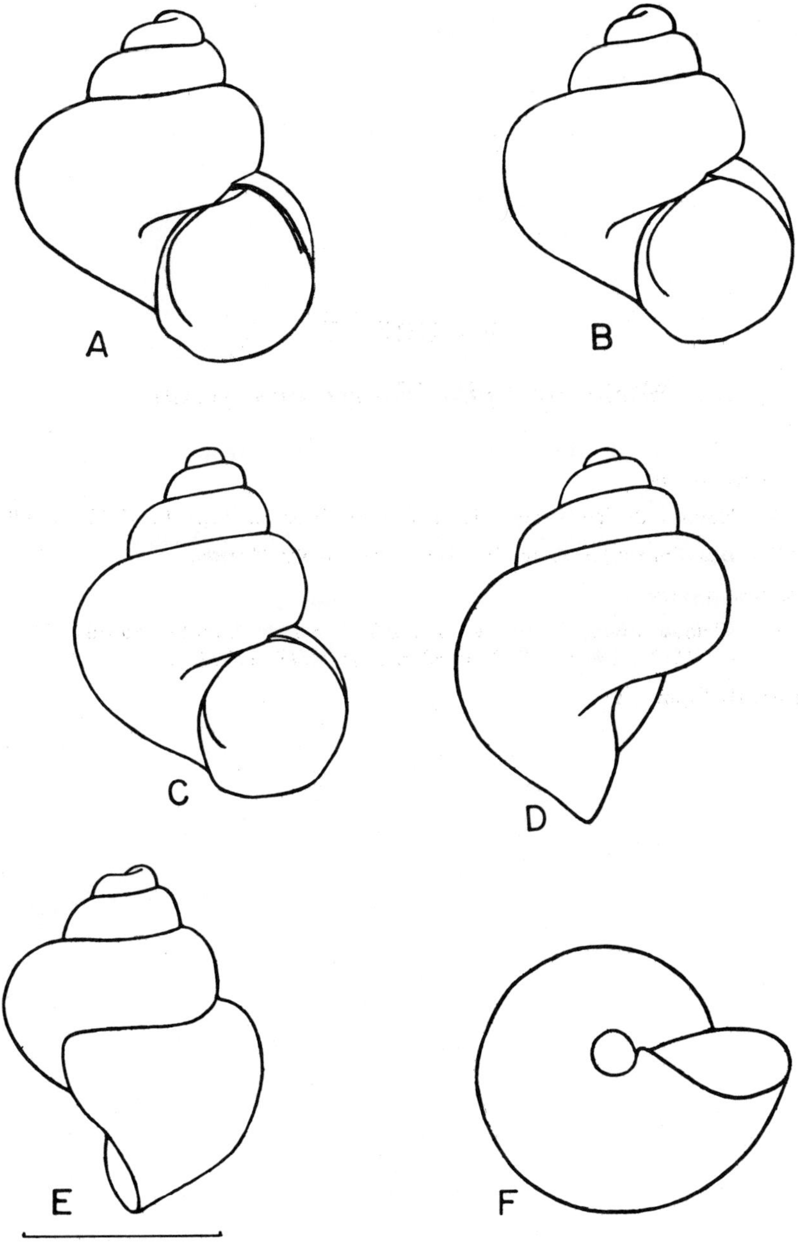

FIGURE 57

Shells of *Spilochlamys* new genus

S. *conica* new species

A. Natural Bridge Springs, Leon County, Florida. Type (*UMMZ* 216480).

B,C. Blue Springs, 3 mi. w. Bronson, Levy County, Florida.

S. *gravis* new species

D–F. Wekiva River, 9 mi. w. Sanford, Seminole County, Florida. D. Type (*UMMZ* 216482). E,F. Paratypes (*UMMZ* 216483).

Scale equals 1 mm.

FIGURE 57

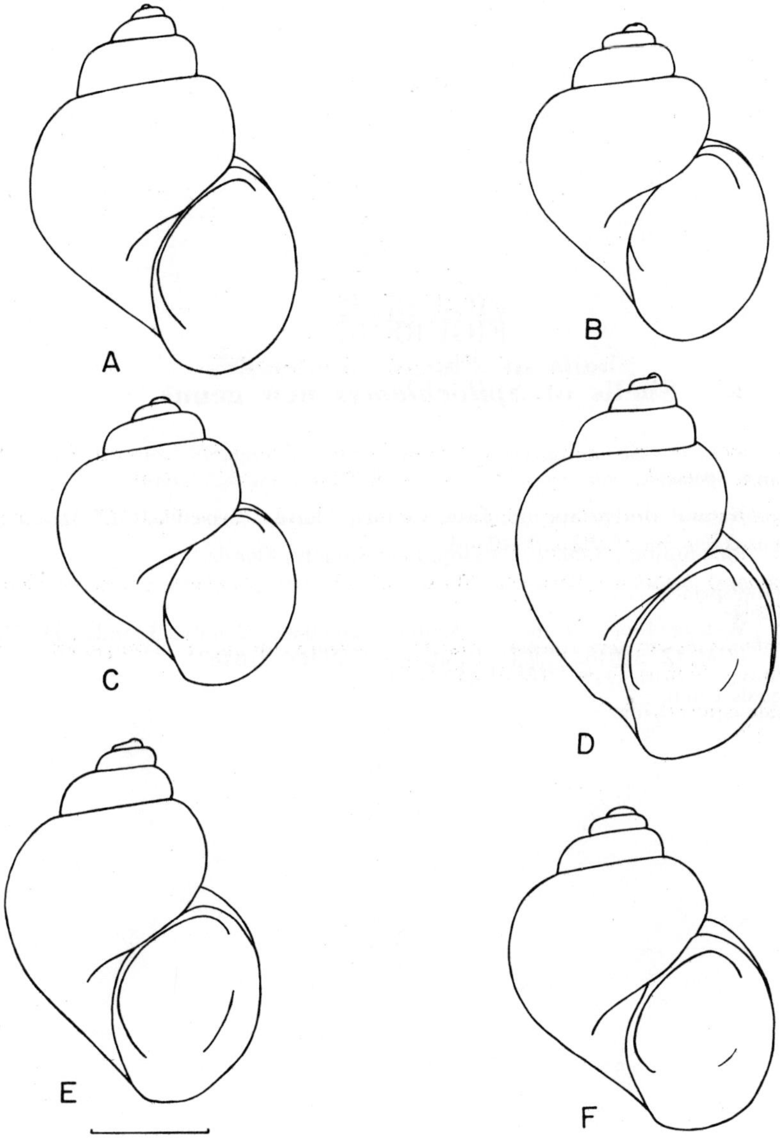

FIGURE 58

Shells of Florida hydrobiids

A,B. *Heleobops docima* new species. Nine-Mile Pond, Everglades National Park, Dade County, Florida. Paratypes: A. Female; B. Male (*UMMZ* 216491).

C. *Hyalopyrgus brevissimus* (Pilsbry). Female. Lake Panasoffkee, 3.7 mi. n.n.w. Panasoffkee, Sumter County, Florida.

D. *Onobops jacksoni* (Bartsch). Marsh 3.2 mi. s.w. Yankeetown, Levy County, Florida.

E. *Onobops crassa* new species. Mangrove swamp 1.9 mi. n. Everglades, Collier County, Florida. Type (*UMMZ* 216537).

Each scale equals 1 mm.

FIGURE 58

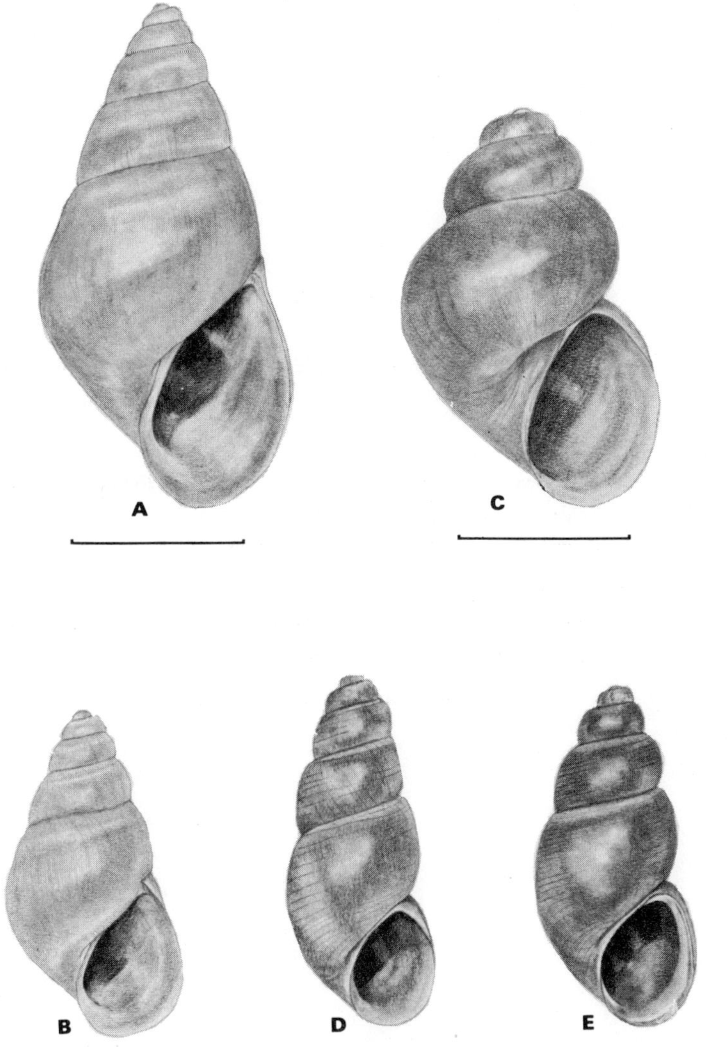

FIGURE 59

Shells of *Pyrgophorus* and *Hyalopyrgus* new genus

P. platyrachis new species

A,B. Canal 7.4 mi. w.n.w. Jupiter, Palm Beach County, Florida. A. Paratype (*UMMZ* 216486). B. Type (*UMMZ* 216485).

C. Pond 0.5 mi. s. Jupiter, Palm Beach County, Florida.

H. aequicostatus (Pilsbry). Female shells.

D. Lake Panasoffkee, 3.7 mi. n.n.w. Panasoffkee, Sumter County, Florida.

E. Creek 7.4 mi. s.e. Geneva, Seminole County, Florida.

F. Blue Springs, 3 mi. w. Orange, Volusia County, Florida.

Scale equals 1 mm.

FIGURE 59

FIGURE 60

Shells of *Littoridinops*

A,B. *L. tenuipes* (Couper). Kitchen Creek, Jonathan Dickinson State Park, Martin County, Florida.

C,D. *L. palustris* new species. Marsh 3.2 mi. s.w. Yankeetown, Levy County, Florida, Type (*UMMZ* 216492).

Scale equals 1 mm.

FIGURE 60

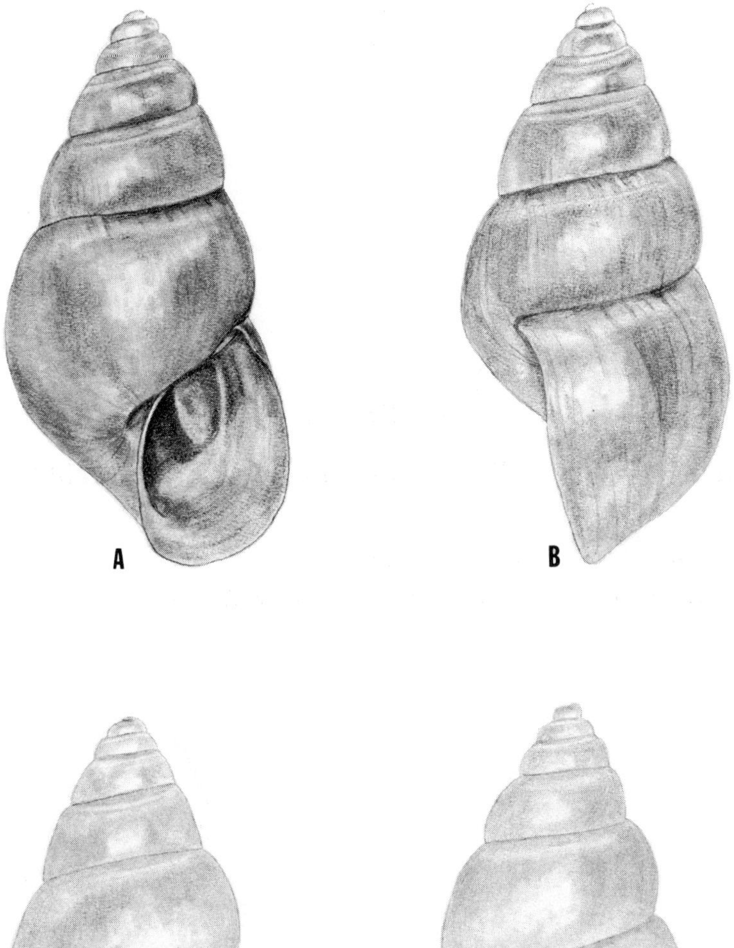

FIGURE 61

Shells of *Littoridinops monroensis* (Frauenfeld)

A,B. St. Johns River, Astor, Lake County, Florida.

C,D. West Lake, Everglades National Park, Dade County, Florida.

Scale equals 1 mm.

FIGURE 61

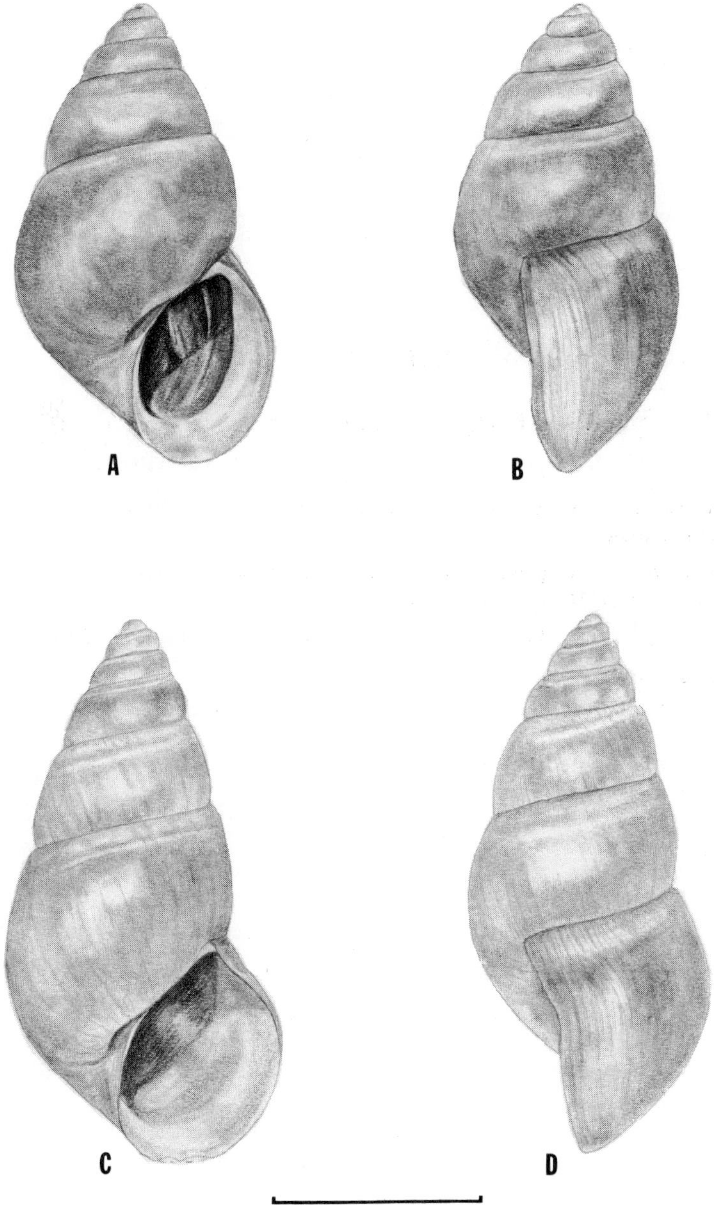

FIGURE 62

Shells of *Aphaostracon* new genus

A,B. *A. rhadinus* new species. Fish Creek, Putnam County, Florida. Type (*UMMZ* 216525), female.

C,D. *A. hypohyalina* new species. Mill Spring, Ichetucknee Springs, Columbia County, Florida. Type (*UMMZ* 216528), female.

E,F. *A. monas* (Pilsbry). Wekiwa Springs, Orange County, Florida. Female.

Scale equals 1 mm.

FIGURE 62

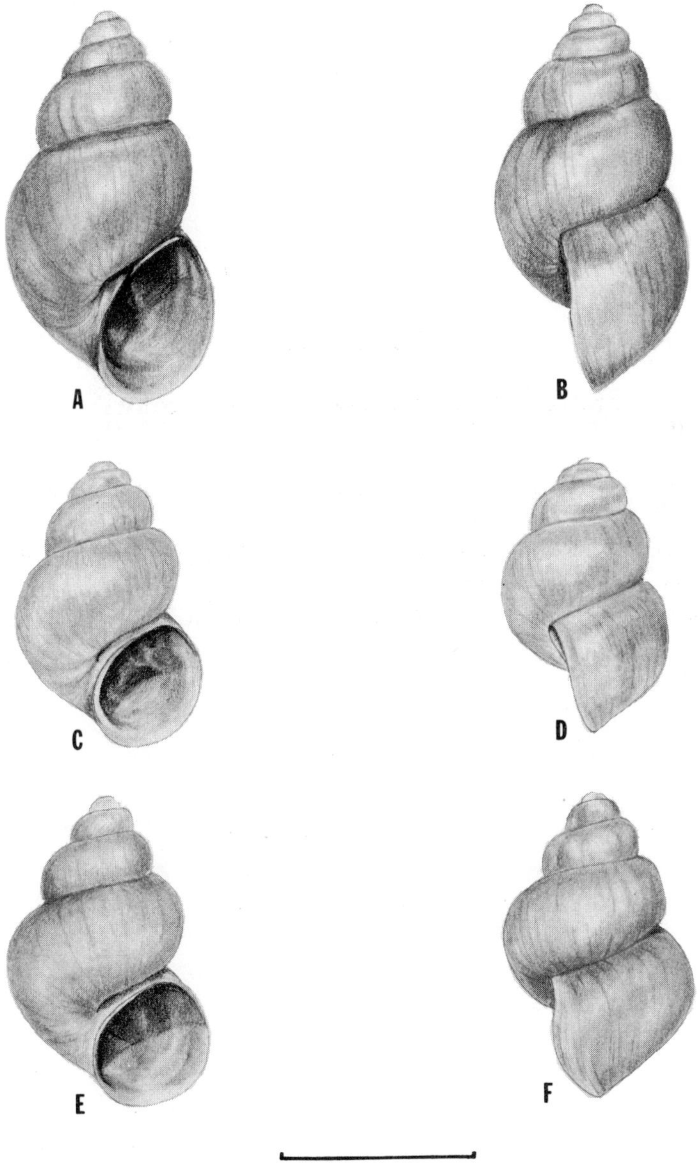

FIGURE 63

Shells of *Aphaostracon* new genus

A,B. *A. pycnus* new species. Alexander Spring Run, Lake County, Florida. Type (*UMMZ* 216512), female.

C,D. *A. asthenes* new species. Blue Springs, 3 mi. w. Orange, Volusia County, Florida. Type (*UMMZ* 216519), female.

E,F. *A. theiocrenetus* new species. Clifton Springs, Seminole County, Florida. Type (*UMMZ* 216510), female.

Scale equals 1 mm.

FIGURE 63

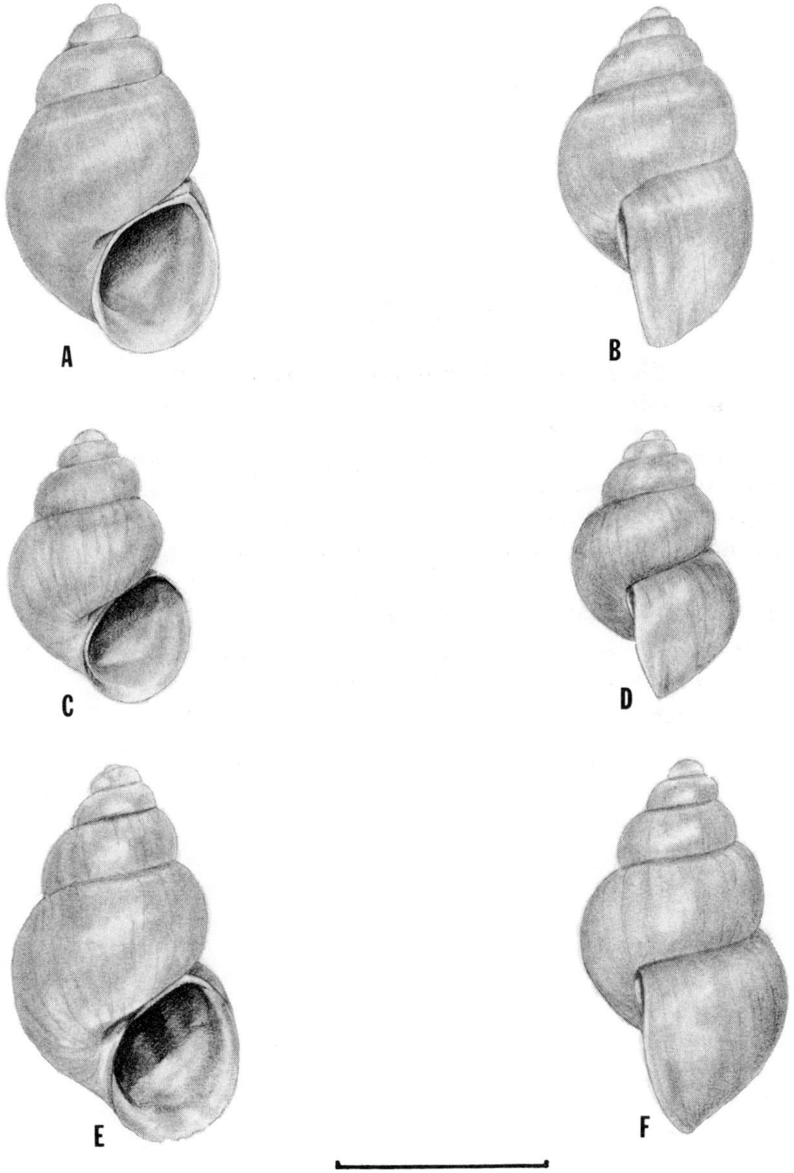

FIGURE 64

Shells of *Aphaostracon* new genus

A,B. *A. pachynotus* new species. Canal 7.4 mi. w. Jupiter, Palm Beach County, Florida. Type (*UMMZ* 216521), female.

C. *A. pachynotus* new species. Canal along U.S. Highway 27, 6 mi. n. Dade County line, Broward County, Florida. (*UMMZ* 216524), male.

D,E. *A. xynoelictus* new species. Fenney Springs, 2 mi. e. Coleman, Sumter County, Florida. Type (*UMMZ* 216517), female.

F,G. *A. chalarogyrus* new species. Magnesia Springs, 3.7 mi. w. Hawthorne, Alachua County, Florida. Type (*UMMZ* 216515), female.

Scale equals 1 mm.

FIGURE 64

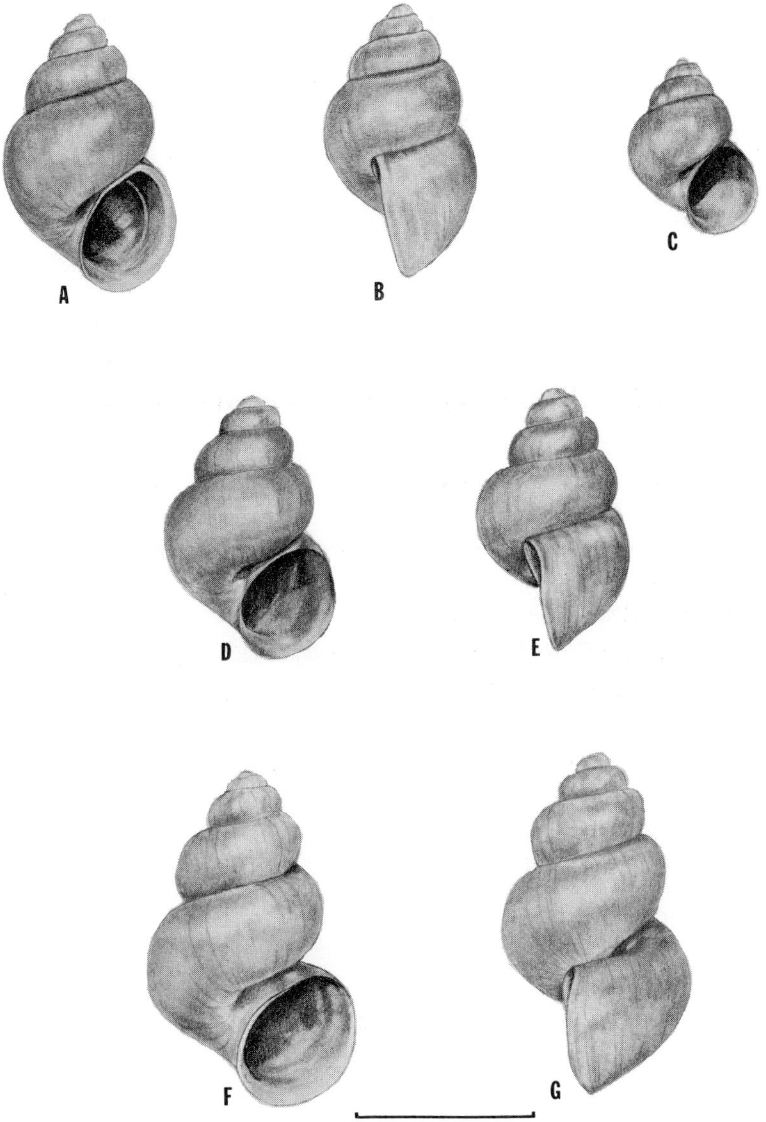

FIGURE 65

Shells of *Spilochlamys* new genus and *Notogillia*

A,B. S. *gravis* new species. Wekiva River, 9 mi. w. Sanford, Seminole County, Florida. Type (*UMMZ* 216482).

C. S. *conica* new species. Natural Bridge Springs, Leon County, Florida. Type (*UMMZ* 216480).

D. *N. wetherbyi* (Dall). Lake Eustis, Lake County, Florida.

E. *N. wetherbyi* (Dall). Rock Springs, Orange County, Florida.

Each scale equals 1 mm.

FIGURE 65

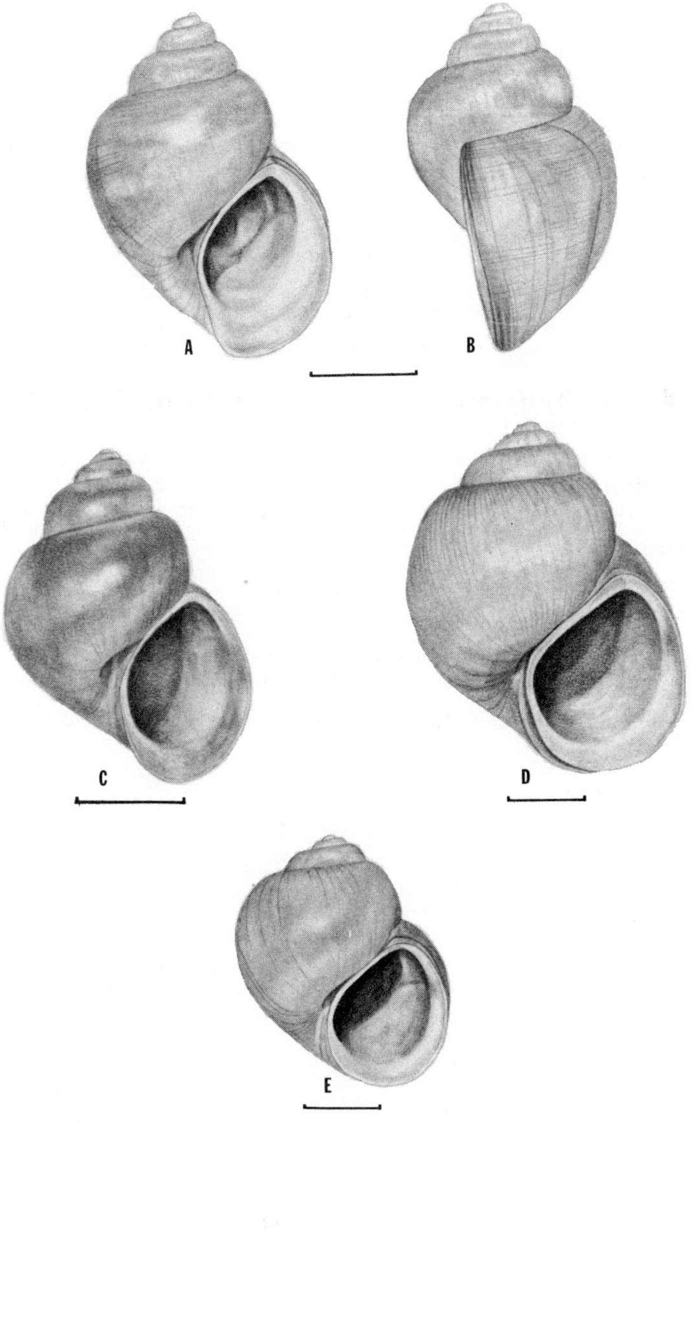

FIGURE 66

Shells of *Cincinnatia*

A. *C. monroensis* (Dall). Benson's Mineral Springs, Enterprise, Volusia County, Florida. "Paratype" (University of Miami collection, ex. C. T. Simpson Collection).

B. *C. monroensis* (Dall). Seepage run 0.5 mi. n.w. Enterprise, Volusia County, Florida. Collected April 21, 1962.

C. *C. floridana* (Frauenfeld). Spring 2.3 mi. w. High Springs, Alachua County, Florida. Lentic form.

D. *C. floridana* (Frauenfeld). Withlacoochee River, Citrus County, Florida. Lotic form.

E. *C. mica* new species. Small spring 1 mi. n.e. U.S. Highway 27, Ichetucknee Springs, Suwannee County, Florida. Type (*UMMZ* 216494).

F. *C. parva* new species. Blue Springs, 3 mi. w. Orange, Volusia County, Florida. Type (*UMMZ* 216530).

Scale equals 1 mm.

FIGURE 66

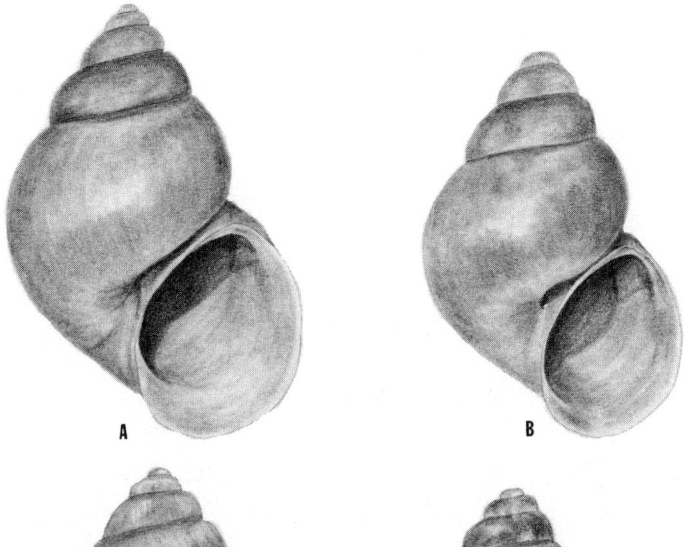

FIGURE 67

Shells of *Cincinnatia*

A. *C. vanhyningi* (Vanatta). Seminole Springs, Lake County, Florida.

B. *C. wekiwae* new species. Wekiwa Springs, Orange County, Florida. Type (*UMMZ* 216503).

C. *C. ponderosa* new species. Sanlando Springs, Seminole County, Florida. Type (*UMMZ* 216499).

D. *C. petrifons* new species. Rock Springs, Orange County, Florida. Type (*UMMZ* 216501).

Scale equals 1 mm.

FIGURE 67

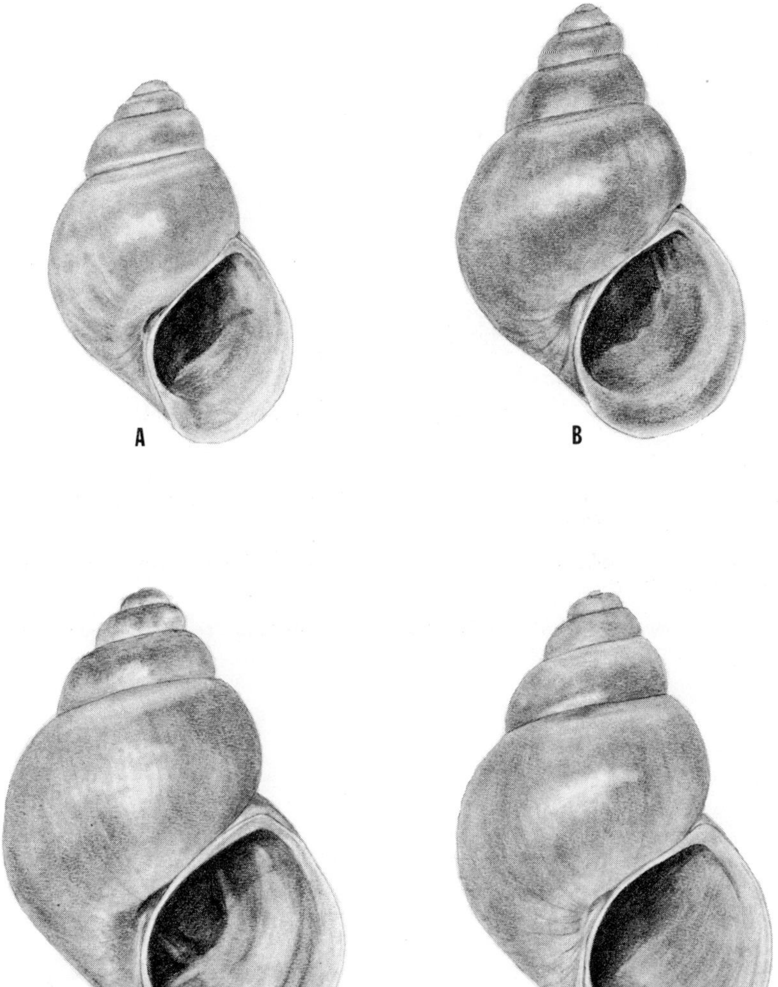

FIGURE 68

Shells of *Amnicola* and *Cincinnatia*

A,B. *A. retromargo* new species. Rice Creek, 7.3 mi. w.n.w. Palatka, Putnam County, Florida. Type (*UMMZ* 216507).

C. *C. fraterna* new species. Creek 4 mi. s. Tocoi, St. Johns County, Florida. Type (*UMMZ* 216532).

D. *C. helicogyra* new species. Crystal Springs, Citrus County, Florida. Type (*UMMZ* 216496).

Scale equals 1 mm.

FIGURE 68

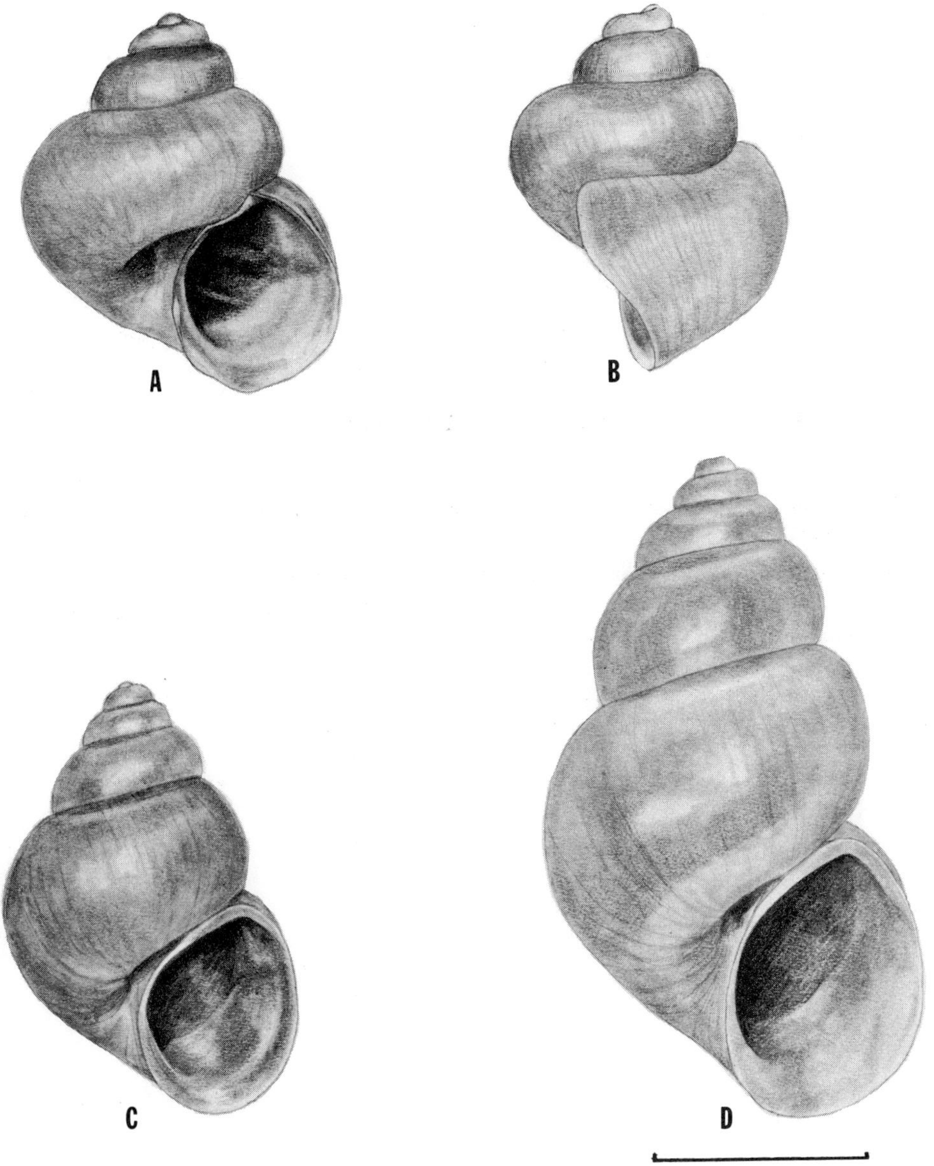

FIGURE 69

Shells of *Amnicola*

A,B. *A. dalli johnsoni* Pilsbry. Heilbronn Springs, 5.6 mi. n.n.w. Starke, Bradford County, Florida.

C,D. *A. d. dalli* (Pilsbry). Wekiwa Springs, Orange County, Florida.

E,F. *A. rhombostoma* new species. Peters Creek, 3.5 mi. n.n.w. Green Cove Springs, Clay County, Florida. Type (*UMMZ* 216505).

Each scale equals 1 mm.

FIGURE 69

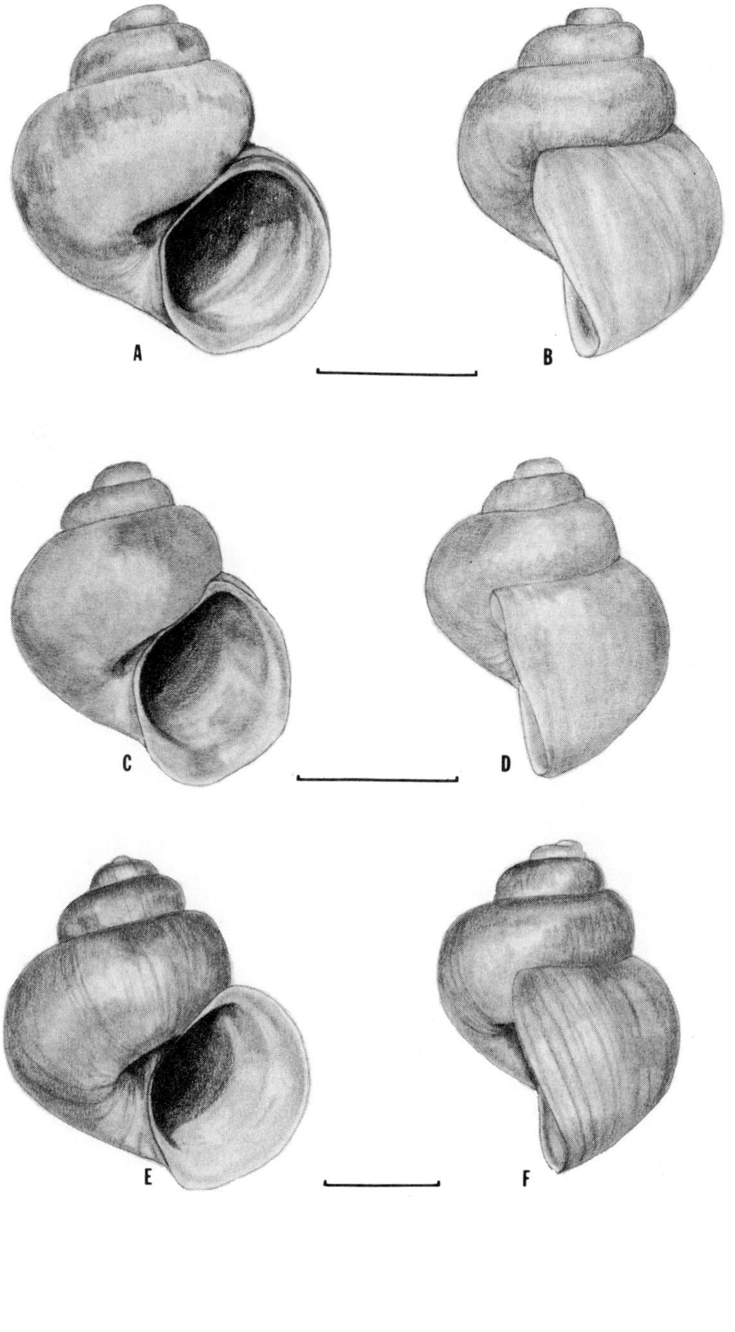

Index